Quantitative Methods for
Current Environmental Issues

Springer

London
Berlin
Heidelberg
New York
Barcelona
Hong Kong
Milan
Paris
Singapore
Tokyo

Clive W. Anderson, Vic Barnett, Philip C. Chatwin
and Abdel H. El-Shaarawi (Eds)

Quantitative Methods for Current Environmental Issues

With 104 Figures

Springer

Clive W. Anderson, MA, MSc, PhD
School of Mathematics and Statistics, University of Sheffield, The Hicks Building,
Sheffield S3 7RH, UK

Vic Barnett, MSc, PhD, DSc, FIS
School of Mathematical Sciences, University of Nottingham, University Park,
Nottingham NG7 2RD, UK

Philip C. Chatwin, MA, PhD, FSS
School of Mathematics and Statistics, University of Sheffield, The Hicks Building,
Sheffield S3 7RH, UK

Abdel H. El-Shaarawi, PhD
National Water Research Institute, 867 Lakeshore Road, Burlington,
Ontario L7R 4A6, Canada

British Library Cataloguing in Publication Data
Quantitative methods for current environmental issues
    1. Environmental sciences – Statistical methods – Congresses
    2. Environmental monitoring – Statistical methods – Congresses
    I. Anderson, C.W. (Clive William), 1944-
    333.7'015195
ISBN-13: 978-1-4471-1171-9   e-ISBN-13: 978-1-4471-0657-9
DOI: 10.1007/978-1-4471-0657-9
Library of Congress Cataloging-in-Publication Data
Quantitative methods for current environmental issues / Clive Anderson ... [et al.], eds.
        p. cm.
    Includes bibliographical references and index.
    ISBN-13: 978-1-4471-1171-9 (alk. paper)

    1. Environmental sciences—Statistical methods—Congresses. I. Anderson, C.W.
    (Clive W.) II. TIES/SPRUCE 2000 (2000 : University of Sheffield)
GE45.S73 Q36 2002
363.7'007'27—dc21   2001055021

ISBN-13: 978-1-4471-1171-9 Springer-Verlag London Berlin Heidelberg
a member of BertelsmannSpringer Science+Business Media GmbH
http://www.springer.co.uk

Typesetting: Camera-ready by the editors and Thomas Unger

12/3830-543210 SPIN 10761755

Preface

One result of the developing public interest in and concern about environmental issues in recent years has been an increased worldwide emphasis on relevant research. During the 1990s, two organizations were established that fostered this research, particularly its quantitative aspects. They were TIES (The International Environmetrics Society) and SPRUCE (Statistics in Public Resources, Utilities and in Care of the Environment). Amongst their varied activities has been the holding of regular and successful conferences. It was perhaps inevitable that the two organizations would one day hold a joint conference, and the first of these – TIES/SPRUCE 2000 – took place at the University of Sheffield, UK from 4 - 8 September 2000.

The conference was organized to incorporate the most successful practices of previous meetings of both TIES and SPRUCE. It was considered specially important to emphasize the essentially multidisciplinary nature of research into environmental issues. The best quantitative work in the field requires expertise both in the underlying science and in statistical and/or mathematical methods. Such work is becoming more widespread as understanding deepens and new collaborations are formed, but it remains difficult. It is hoped that conferences such as TIES/SPRUCE 2000 can help promote the necessary awareness and communication of ideas fundamental to such work, and can add something to the experience of participants which will place them in a stronger position to emphasize the need for new combinations of skills in the training of young research scientists.

The theme of TIES/SPRUCE 2000 was *Quantitative Methods in Current Environmental Issues.* Eleven distinguished scientists or engineers gave invited plenary lectures. They were:

Julian Besag	*University of Washington, Seattle, USA*
Steve Buckland	*University of St Andrews, UK*
(J Stuart Hunter Lecturer)	
Brad Carlin	*University of Minnesota, USA*
Joe Chang	*George Mason University, Virginia, USA*
Dave Higdon	*Duke University, North Carolina, USA*
Gudmund Host	*Norwegian Computing Centre, Oslo, Norway*
Jim McQuaid	*University of Sheffield, UK*
Tony O'Hagan	*University of Sheffield, UK*
Paul Sampson	*University of Washington, Seattle, USA*
Hans Wackernagel	*Ecole des Mines de Paris, Fontainebleau, France*
Lucy Wyatt	*University of Sheffield, UK*

This book contains articles based on nine of these presentations, together with three others based on peer-reviewed contributed papers (of which there were more than 70). The articles are grouped around the themes of

Spatial and Temporal Models and Methods,
Environmental Sampling and Standards,
Atmosphere and Ocean, and
Risk and Uncertainty.

The success of TIES/SPRUCE 2000 owed much to the enthusiasm and hard work of the Organizing Committee. The Co-Chairs (Clive Anderson and Philip Chatwin) wish to record their thanks to Nils Mole, above all, and also to Paul Blackwell, Nirvana Bloor, Nancy Doss, Marianna Keray, Richard Martin, Maged Messeh, Rick Munro, Ruth Parker and Robertus von Fay-Siebenburgen.

The present volume contains a selection of the most important work now being undertaken on quantitative aspects of environmental issues. We hope it will prove interesting and valuable.

Clive Anderson
Vic Barnett
Philip Chatwin
Abdel El-Shaarawi

Sheffield
Nottingham
Burlington

Contents

Part IV. Risk and Uncertainty

## Part I

Spatial and Temporal Models and Methods

1 Modeling Spatio-Temporally Misaligned Areal and Point Process Environmental Data

Bradley P. Carlin[1], Andrew S. Mugglin[2], Li Zhu[3], and Alan E. Gelfand[4]

[1] Division of Biostatistics, University of Minnesota,
420 Delaware St. SE, Minneapolis, MN 55455, USA
[2] Clinical Outcomes Research, Cardiac Rhythm Management,
Medtronic Inc., 4000 Lexington Avenue N., MN 55126, USA
[3] Department of Epidemiology & Biostatistics, and Department of Statistics,
Texas A&M University, College Station TX 77843, USA
[4] Department of Statistics, University of Connecticut,
Box U-120, Storrs CT 06269-3120, USA

Abstract. We consider inference using multivariate data that are misaligned both in space and in support. We begin by considering the analysis of random variables (typically counts or rates) which are aggregated over differing sets of possibly non-nested regional boundaries. This sort of *areal misalignment* is handled using conditionally independent Poisson-multinomial models, thus offering a Bayesian solution to the celebrated *modifiable areal unit problem* (MAUP). Explanatory covariates and multilevel responses can also be easily accommodated, with spatial correlation modeled using a conditionally autoregressive (CAR) prior structure. This approach is illustrated with a dataset on area-level population counts near a putative radiation source in southwestern Ohio. We then turn to a more general setting where we assume an underlying continuous spatial process $Y(s)$ for locations $s \in D$, a region of interest. This allows us to consider the general *change of support problem* (COSP), where we seek to make inferences about the values of a variable at either points *or* regions different from those at which it has been observed. Next, we apply our COSP approach to the *spatio-temporal* case, and show that the additional computational burden to analyze the correspondingly larger data set still emerges as manageable. Here we illustrate using a dataset of observed ozone levels in the Atlanta, Georgia metropolitan area. As with many recent hierarchical Bayesian spatial applications, computing is implemented via carefully tailored Metropolis-Hastings algorithms, with map summaries created using a geographic information system (GIS).[5]

1.1   Introduction

Consider a univariate variable that is spatially observed. In particular, assume that it is observed either at points in space, which we refer to as *point-referenced* or simply *point* data, or over areal units (e.g., counties or zip codes), which we refer to as *block* data. The *change of support problem* is concerned with inference about the values of the variable at points or blocks different from those at which it has been observed.

In the case where the data are collected exclusively at blocks (*source* zones) and inference is sought exclusively at new blocks (*target* zones), the resulting areal interpolation problem is often referred to as the *modifiable areal unit problem* (see e.g. Cressie 1996), and has a long history in the geography literature; see e.g. Goodchild & Lam (1980), Lam (1983), Flowerdew & Green (1989, 1991, 1992, 1994), Flowerdew *et al.* (1991), Langford *et al.* (1991), Goodchild *et al.* (1993), and Fisher & Langford (1995). In the case of an *extensive* variable (i.e., one whose value for a block can be viewed as a sum of sub-block values, as in the case of population, disease counts, productivity or wealth), the simplest and most common approach is *areal interpolation*. This is the process by which source values are imputed to target zones proportionally to the areas of the zones. This method is widely available in many GISs, but obviously depends on the unrealistic assumption that the source values are uniformly distributed over the entire area in question.

In the statistical literature, Tobler (1979) offers a block-block interpolation approach that goes beyond mere areal weighting. More recently, Mugglin & Carlin (1998) and Mugglin *et al.* (1999) present a hierarchical Bayesian method for interpolation and smoothing of Poisson responses with covariates, where the principal advantage is full inference (e.g., enabling interval estimates) for the distributions of target zone populations. In their setting the source zones are U.S. census tracts, while the target zones (and the zones on which covariate data are available) are U.S. census block groups. That is, the target zonation is merely a refinement of the source zonation, a situation they term *nested* misalignment. Zhu *et al.* (2000) extend this approach to the spatio-temporal case, accommodating nested misalignment within time periods, and also temporal misalignment arising from changes in the regional boundaries over time. Best *et al.* (2000) propose a method of Bayesian estimation for Gamma/Poisson random field models, in which the data are misaligned and not nested. In their setting the response regions are small enough that, together with the covariate values, they can be regarded as a *marked point process*.

As mentioned above, the change of support problem goes beyond areal interpolation (changing regional boundaries) to include changing the type of support as well (say, from point sources to target zones). The case of point-point misalignment and prediction has historically been referred to as *kriging* (see e.g. Cressie 1993). Bayesian approaches to kriging date to the Bayesian nonparametric Gaussian process modeling of O'Hagan (1978); in his discussion Beale (1978) points out the connection to kriging. More recent work on Bayesian spatial prediction includes Le & Zidek (1992), Handcock & Stein (1993), Brown, Le & Zidek (1994), Handcock & Wallis (1994), DeOliveira, Kedem & Short (1997), Ecker & Gelfand (1997), Diggle, Tawn & Moyeed (1998), and Karson *et al.* (1999). As in other contexts, use of the Bayesian paradigm in kriging enables full inference (a posterior predictive distribution for every prediction of interest, joint distributions for all pairs of predictions, etc.) and avoiding asymptotics. Still, in our work we typically use rather vague prior distributions, so that our results will roughly resemble those of a likelihood analysis.

A way of handling completely general support is to assume the existence of an underlying spatial process taking the value $Y(s)$ at each location s. Point-level data then consists of $Y(s_i), i = 1, \ldots, I$, while block-level data (over some spatial block B) are assumed to arise as block averages, i.e. $Y(B) = |B|^{-1} \int_B Y(s)ds$, where $|B|$ denotes the area of block B. Inference about blocks through averages is not only formally attractive but demonstrably preferable to *ad hoc* approaches. One such approach would be to average over the observed $Y(s_i)$ in B. But this presumes there is at least one observation in any B, and ignores the information about the spatial process in the observations outside of B. Another *ad hoc* approach would be to simply predict the value at some central point of B. But this value has larger variability than (and may be biased for) the block average.

The remainder of this paper is organized as follows. Section 1.2 describes a framework for hierarchical Bayesian interpolation, estimation, and spatial smoothing over *non-nested* misaligned data grids in settings where the regions are too large to be regarded as marked point processes. Section 1.3 then illustrates the approach with an analysis of a dataset collected in response to possible radon contamination resulting from the former Feed Materials Production Center (FMPC) in southwestern Ohio. Section 1.4 takes up the more general point-block case, first considering methodology for spatial data at a single time point and then tackling a more general spatio-temporal model. Section 1.5 applies these approaches to a dataset of point-level ozone measurements in greater Atlanta, Georgia, where the goal is to obtain zip code-level block averages. Finally, in Section 1.6 we assess our findings and suggest avenues for further research.

1.2 Misaligned Areal Data Model Development

We confine our model development to the case of two misaligned spatial grids. Given this development, the extension to more than two grids will be conceptually apparent. The additional computational complexity and bookkeeping detail will also be evident.

Let the first grid have regions indexed by $i = 1, \ldots, I$, denoted by B_i, and let $S_B = \bigcup_i B_i$. Similarly, for the second grid we have regions C_j, $j = 1, \ldots, J$

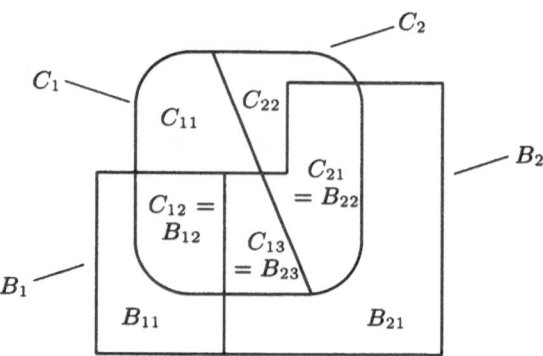

Fig. 1.1. Illustrative representation of areal data misalignment

with $S_C = \bigcup_j C_j$. In some applications $S_B = S_C$, i.e., the B-cells and the C-cells offer different partitions of a common region. A special case includes the situation where one partition is a refinement of the other, e.g., the case where each C_j is contained entirely in one and only one B_i (as considered by Mugglin & Carlin 1998). Another possibility is that one data grid contains the other; say, $S_B \subset S_C$. In this case, there will exist some C cells for which a portion lies outside of S_B. In the most general case, there is no containment and there will exist B-cells for which a portion lies outside of S_C and C-cells for which a portion lies outside of S_B. Figure 1.1 illustrates this most general situation.

Atoms are created by intersecting the two grids. For a given B_i, each C-cell which intersects B_i creates an atom (which possibly could be a union of disjoint regions). There may also be a portion of B_i which does not intersect with any C_j. We refer to this portion as the *edge* atom associated with B_i, i.e., a B-edge atom. In Figure 1.1, atoms B_{11} and B_{21} are B-edge atoms. Similarly, for a given C_j, each B-cell which intersects with C_j creates an atom, and we analogously determine C-edge atoms (atoms C_{11} and C_{22} in Figure 1.1). It is crucial to note that each non-edge atom can be referenced relative to an appropriate B-cell, say B_i, and denoted as B_{ik}. It also can be referenced relative to an appropriate C cell, say C_j, and denoted by $C_{j\ell}$. Hence, there is a one-to-one mapping within $S_B \bigcap S_C$ between the set of ik's and the set of $j\ell$'s, as shown in Figure 1.1 (which also illustrates our convention of indexing atoms by area, in descending order). Formally we can define the function c on non-edge B-atoms such that $c(B_{ik}) = C_{j\ell}$, and the *inverse* function b on C-atoms such that $b(C_{j\ell}) = B_{ik}$. For computational purposes we suggest creation of "look-up" tables to specify these functions. (Note that the possible presence of both types of edge cell precludes a single "ij" atom numbering system, since such a system could index cells on either S_B or S_C, but not their union.)

Without loss of generality we refer to the first grid as the *response* grid, that is, at each B_i we observe a response Y_i. We seek to explain Y_i using a variety of covariates. Some of these covariates may, in fact, be observed on the response grid; we denote the value of this vector for B_i by $\boldsymbol{W}_i$. But also, some covariates are observed on the second or *explanatory* grid. We denote the value of this vector for C_j by $\boldsymbol{X}_j$.

We seek to explain the observed Y's through both $\boldsymbol{X}$ and $\boldsymbol{W}$. The misalignment between the $\boldsymbol{X}$'s and Y's is the obstacle to standard regression methods. What levels of $\boldsymbol{X}$ should be assigned to Y_i? We propose a fully model-based approach in the case where the Y's and X's are aggregated measurements. The advantage of a model-based approach implemented within a Bayesian framework is full inference both with regard to estimation of model parameters and prediction using the model.

The assumption that the Y's are aggregated measurements means Y_i can be envisioned as $\sum_k Y_{ik}$, where the Y_{ik} are unobserved or latent and the summation is over all atoms (including perhaps an edge atom) associated with B_i. To simplify, we assume that the X's are also scalar aggregated measurements, i.e., $X_j = \sum_\ell X_{j\ell}$ where the summation is over all atoms associated with C_j. As for the $\boldsymbol{W}$'s, we assume that each component is either an aggregated measurement or an *inheritable* measurement. For component r, in the former case $W_i^{(r)} = \sum_k W_{ik}^{(r)}$ as with Y_i; in the latter case $W_{ik}^{(r)} = W_i^{(r)}$. In addition to (or perhaps in place of) the $\boldsymbol{W}_i$ we will introduce B-cell random effects μ_i, $i = 1, ..., I$. These effects are employed to capture spatial association among the Y_i's. The μ_i can be given a spatial prior specification. A Markov random field form (Besag 1974, Bernardinelli & Montomoli 1992), as described below, is convenient. Similarly we will introduce C-cell random effects ω_j, $j = 1, ..., J$ to capture spatial association among the X_j's. It is assumed that the latent Y_{ik} inherit the effect μ_i and that the latent $X_{j\ell}$ inherit the effect ω_j.

For aggregated measurements which are counts, we assume the latent variables are conditionally independent Poissons. As a result, the observed measurements are Poissons as well and the conditional distribution of the latent variables given the observed is a product multinomial. For aggregated measurements which are continuous, a convenient distributional assumption is conditionally independent Gammas whence the latent variables would be rescaled to product Dirichlet. An alternative choice is the normal, whence the latent variables would have a distribution which is a product of conditional multivariate normals. In the sequel we detail the Poisson case.

As mentioned above, area traditionally plays an important role in allocation of spatial measurements. Letting $|A|$ denote the area of region A, if we apply the standard assumption of allocation proportional to area to the $X_{j\ell}$ in a stochastic fashion, we would obtain

$$X_{j\ell} \mid \omega_j \sim Po(e^{\omega_j}|C_{j\ell|}) \,, \tag{1.1}$$

assumed independent for $\ell = 1, 2, ..., L_j$. Then $X_j \mid \omega_j \sim Po(e^{\omega_j}|C_j|)$ and $(X_{j1}, X_{j2}, ..., X_{j,L_j} \mid X_j, \omega_j) \sim \text{Mult}(X_j; q_{j1}, ..., q_{j,L_j})$ where $q_{j\ell} = |C_{j\ell}|/|C_j|$. Such strictly area-based modeling cannot be applied to the Y_{ik}'s since it fails to connect the Y's with the X's (as well as the W's). To do so we again begin at the atom level. For non-edge atoms we use the previously mentioned look-up table to find the $X_{j\ell}$ to associate with a given Y_{ik}. It is convenient to denote this $X_{j\ell}$ as X'_{ik}. Ignoring the W_i for the moment, we assume

$$Y_{ik} \mid \mu_i, \theta_{ik} \sim Po\left(e^{\mu_i}|B_{ik}| \, h(X'_{ik}/|B_{ik}| \, ; \theta_{ik})\right) \qquad (1.2)$$

independent for $k = 1, 2, ..., K_i$. Here h is a preselected parametric function, the part of the model specification which adjusts an expected proportional-to-area allocation according to X'_{ik}. Since (1.1) models expectation for $X_{j\ell}$ proportional to $|C_{j\ell}|$, it is natural to use the *standardized* form $X'_{ik}/|B_{ik}|$ in (1.2). Particular choices of h include $h(z \, ; \theta_{ik}) = z$ yielding $Y_{ik} \mid \mu_i \sim Po(e^{\mu_i}X'_{ik})$, which would be appropriate if we chose not to use $|B_{ik}|$ explicitly in modeling $E(Y_{ik})$. In our FMPC implementation, we actually select $h(z \, ; \theta_{ik}) = z + \theta_{ik}$ where $\theta_{ik} = \theta/(K_i|B_{ik}|)$ and $\theta > 0$; see equation (1.8) below and the associated discussion. If B_i has no associated edge atom, then

$$Y_i \mid \mu_i, \theta, \{X_{j\ell}\} \sim Po\left(e^{\mu_i} \sum_k |B_{ik}| \, h(X'_{ik}/|B_{ik}| \, ; \theta_{ik})\right). \qquad (1.3)$$

If B_i has an edge atom, say B_{iE}, since there is no corresponding $C_{j\ell}$ there is no corresponding X'_{iE}. Hence, we introduce a latent X'_{iE} whose distribution is determined by the non-edge atoms which are neighbors of B_{iE}. Paralleling equation (1.1), we model X'_{iE} as

$$X'_{iE} \mid \omega^*_i \sim Po(e^{\omega^*_i}|B_{iE}|) , \qquad (1.4)$$

introducing an additional set of random effects $\{\omega^*_i\}$ to the existing set $\{\omega_j\}$. These two sets then *jointly* receive a Markov random field prior. Alternatively, we can model

$$X'_{iE} \sim Po\left(|B_{iE}| \left(\sum_{N(B_{iE})} X'_t / \sum_{N(B_{iE})} |B_t|\right) \frac{1}{m_{iE}} \sum_{N(B_{iE})} X'_t/|B_t|\right),$$

where $N(B_{iE})$ is the set of neighbors of B_{iE} and t indexes this set. Effectively, we multiply $|B_{iE}|$ by the overall count per unit area in the neighboring non-edge atoms. While this model is somewhat more data-dependent than the (more model-dependent) one given in (1.4), we remark that it can actually lead to better MCMC convergence due to the improved identifiability in its parameter space: the spatial similarity of the structures in the edge zones is being modeled directly, rather than indirectly via the similarity of the ω^*_i and the ω_j.

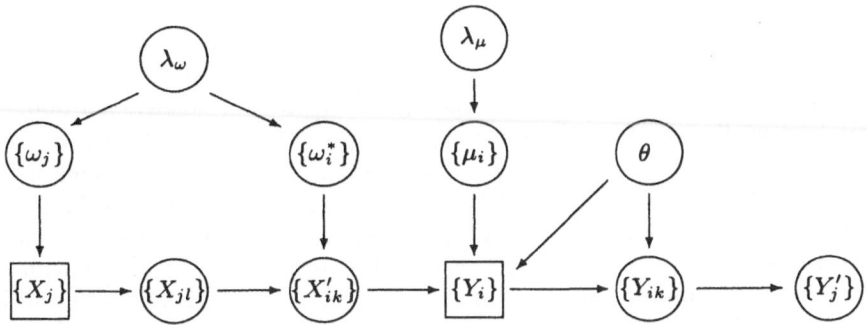

Fig. 1.2. Graphical version of the model, with variables as described in the text. Boxes indicate data nodes, while circles indicate unknowns.

Now, with an X'_{ik} for all ik, (1.2) is extended to all B-atoms and the conditional distribution of Y_i is determined for all i as in (1.3). But also $Y_{i1}, ..., Y_{ik_i} \mid Y_i, \mu_i, \theta_{ik} \sim \text{Mult}(Y_i; q_{i1}, ..., q_{ik_i})$ where the chance of falling in the kth bin is $q_{ik} = |B_{ik}| h(X'_{ik}/|B_{ik}|; \theta_{ik})/\sum_k |B_{ik}| h(X'_{ik}/|B_{ik}|; \theta_{ik})$.

To capture the spatial nature of the B_i we may adopt a Markov random field prior for the μ_i's. Following Bernardinelli & Montomoli (1992) we assume that

$$f(\mu_i \mid \mu_{i',i'\neq i}) = N\left(\sum_{i'} u_{ii'}\mu_{i'}/u_{i.} , 1/(\lambda_\mu u_{i.})\right) \qquad (1.5)$$

where $u_{ii} = 0$, $u_{ii'} = u_{i'i}$ and $u_{i.} = \sum_{i'} u_{ii'}$. The joint distribution associated with these conditional distributions is uniquely determined but improper. Spatial heterogeneity assumes that the physical location of the units determine the $u_{ii'}$. If we put $u_{ii'} = 1$ for $B_{i'}$ a neighbor of B_i and $u_{ii'} = 0$ otherwise, we obtain the standard "adjacency" form of the conditionally autoregressive (CAR) prior (Besag *et al.*, 1995). If, for instance, all $u_{ii'} = 1$ we obtain an exchangeable prior (EXCH) which captures heterogeneity but not local clustering.

Similarly we assume that

$$f(\omega_j \mid \omega_{j',j'\neq j}) = N\left(\sum_{j'} v_{jj'}\omega_{j'}/v_{j.} , 1/(\lambda_\omega v_{j.})\right) . \qquad (1.6)$$

We adopt a proper Gamma prior for λ_μ and also for λ_ω. When θ is present we require a prior which we denote by $f(\theta)$. The choice of $f(\theta)$ will likely be vague but its form depends upon the adopted parametric form of h.

The entire specification can be given a representation as a graphical model, as in Figure 1.2. In this model the arrow from $\{X_{j\ell}\} \to \{X'_{ik}\}$ indicates the inversion of the $\{X_{j\ell}\}$ to $\{X'_{ik}\}$, augmented by any required edge atom values

X'_{iE}. The $\{\omega^*_i\}$ would be generated if the X'_{iE} are modeled using (1.4). Since the $\{Y_{ik}\}$ are not observed, but are distributed as multinomial given the fixed block group totals $\{Y_i\}$, this is a predictive step in our model, as indicated by the arrow from $\{Y_i\}$ to $\{Y_{ik}\}$ in the figure. In fact, as mentioned above the further predictive step to impute Y'_j, the Y total associated with X_j in the jth target zone, is of key interest. If there are edge atoms C_{jE}, this will require a model for the associated Y'_{jE}. Since there is no corresponding B-atom for C_{jE} a specification such as (1.2) is not appropriate. Rather, we can imitate the above modeling for X'_{iE} using (1.4) by introducing $\{\mu^*_j\}$ which along with the μ_i follow the prior in (1.5). The $\{\mu^*_j\}$ and $\{Y'_{jE}\}$ would add two consecutive nodes to the right side of Figure 1.2, connecting from λ_μ to $\{Y'_j\}$.

The entire distributional specification overlaid on this graphical model has been supplied in the foregoing discussion and (in the absence of C_{jE} edge atoms, as in our Section 1.3 example) takes the form

$$
\begin{aligned}
&\prod_i f(Y_{i1}, ..., Y_{ik_i} \mid Y_i, \theta)\, f(\theta) \\
&\quad \times \prod_i f(Y_i \mid \mu_i, \theta, \{X'_{ik}\})\, f(\{X'_{ik}\} \mid \omega^*_i, \{X_{j\ell}\}) \\
&\quad \times \prod_j f(X_{j1}, ..., X_{jL_j} \mid X_j) \prod_j f(X_j \mid \omega_j) \\
&\quad \times f(\{\mu_i\} \mid \lambda_\mu)\, f(\lambda_\mu)\, f(\{\omega_j\}, \{\omega^*_i\} \mid \lambda_\omega)\, f(\lambda_\omega) \,.
\end{aligned}
\tag{1.7}
$$

Bringing in the $\boldsymbol{W}_i$ merely revises the exponential term in (1.2) from $\exp(\mu_i)$ to $\exp(\mu_i + \boldsymbol{W}^T_{ik}\beta)$. Again, for an inherited component of $\boldsymbol{W}_i$, say $W^{(r)}_i$, the resulting $W^{(r)}_{ik} = W^{(r)}_i$. For an aggregated component of $\boldsymbol{W}_i$, again say $W^{(r)}_i$, we imitate (1.1) assuming $W^{(r)}_{ik} \mid \mu^{(r)}_i \sim Po(e^{\mu^{(r)}_i}|B_{ik}|)$, independent for $k = 1, ..., K_i$. A spatial prior on the $\mu^{(r)}_i$ and a Gaussian (or perhaps flat) prior on β completes the model specification.

Finally, on the response grid, for each B_i rather than observing a single Y_i we may observe Y_{im}, where $m = 1, 2, ..., M$ indexes levels of factors such as sex, race or age group. Here we seek to use these factors, in an ANOVA fashion, along with the X_j (and $\boldsymbol{W}_i$) to explain the Y_{im}. Ignoring $\boldsymbol{W}_i$, the resultant change in (1.2) is that Y_{ikm} will be Poisson with μ_i replaced by μ_{im}, where μ_{im} has an appropriate ANOVA form. For example, in the case of sex and age classes, we might have a sex main effect, an age main effect, and a sex-age interaction effect. In our application these effects are not nested within i; we include only a spatial overall mean effect indexed by i. We have not explored nested effects nor have we considered ANOVA forms for the θs (perhaps with an appropriate link function). Modification to incorporate the $\boldsymbol{W}_i$ is straightforward.

1.3   Example: Radon Exposure near an Ohio Contamination Source

1.3.1   Description of Dataset

Risk-based decision making is often used for prioritizing cleanup efforts at U.S. Superfund sites. Often these decisions will be based on estimates of the past, present, and future potential health impacts. These impact assessments usually rely on estimation of the number of outcomes, and the accuracy of these estimates will depend heavily on the ability to estimate the number of individuals at risk. Our motivating dataset is connected with just this sort of risk assessment.

In the years 1951-1988 near the town of Ross in southwestern Ohio, the former Feed Materials Production Center (FMPC) processed uranium for weapons production. Draft results of the Fernald Dosimetry Reconstruction Project, sponsored by the Centers for Disease Control and Prevention (CDC), indicated that during production years the FMPC released radioactive materials (primarily radon and its decay products and, to a lesser extent, uranium and thorium) from the site. Although radioactive liquid wastes were released, the primary exposure to residents of the surrounding community resulted from breathing radon decay products. The potential for increased risk of lung cancer is thus the focus of intense local public interest and ongoing public health studies (see Devine *et al.*, 1998).

Estimating the number of adverse health outcomes in the population (or in subsets thereof) requires estimation of the number of individuals at risk. Population counts, broken down by age and sex, are available from the U.S. Census Bureau according to federal census block groups, while the areas of exposure interest are dictated by both direction and distance from the plant. Rogers & Killough (1997) construct an exposure "windrose," which consists of 10 concentric circular bands at 1-kilometer radial increments divided into 16 compass sectors (N, NNW, NW, WNW, W, etc.). Through the overlay of such a windrose onto U.S. Geological Survey (USGS) maps, they provide counts of the numbers of "structures" (residential buildings, office buildings, industrial building complexes, warehouses, barns, and garages) within each subdivision (*cell*) of the windrose.

Figure 1.3 shows the windrose centered at the FMPC. We assign numbers to the windrose cells, with 1-10 indexing the cells starting at the plant and running due north, then 11-20 running from the plant to the north-northwest, and so on. Structure counts are known for each cell; the hatching pattern in the figure indicates the areal density (structures per square kilometer) in each cell.

Also shown in Figure 1.3 are the boundaries of 39 Census Bureau block groups, for which 1990 population counts are known. These are the source zones for our interpolation problem. Shading intensity indicates the population density (persons per square kilometer) for each block group. The inter-

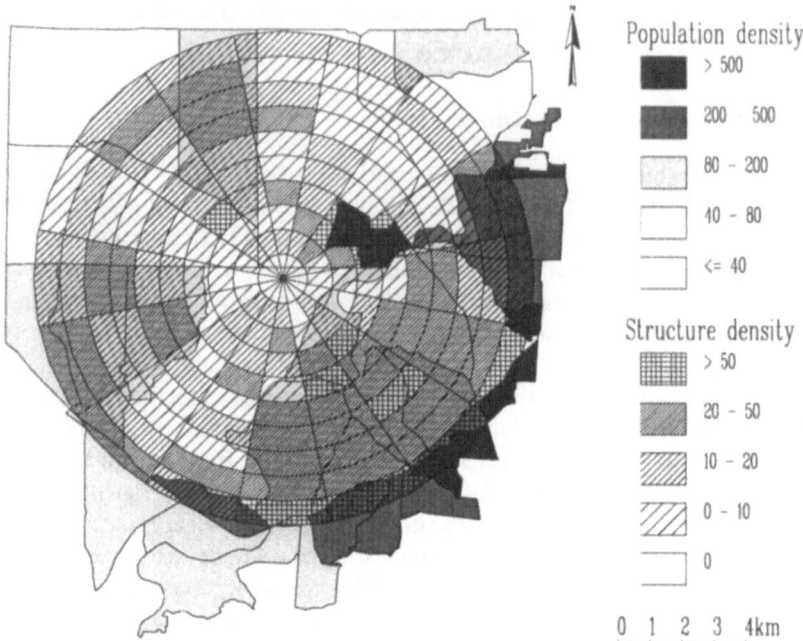

Fig. 1.3. Census block groups and 10-km windrose near the FMPC site, with 1990 population density by block group and 1980 structure density by cell (both in counts per km²).

section of the two (non-nested) zonation systems results in 389 regions we call *atoms*, which can be aggregated appropriately to form either cells or block groups.

The plant was in operation for 38 years, raising concern about the potential health risks it has caused – a question that has been under active investigation by the CDC for some time. Present efforts to assess the impact of the FMPC on cancer morbidity and mortality require the analysis of this misaligned dataset; in particular, it is necessary to interpolate gender- and age group-specific population counts to the windrose exposure cells. These numbers of persons at risk could then be combined with cell-specific dose estimates obtained by Killough *et al.* (1996) and estimates of the cancer risk per unit dose to obtain expected numbers of excess cancer cases by cell. The resulting expected counts would not only suggest the overall level of risk to the community, but also indicate whether sufficient power existed to justify the initiation of a full epidemiological study of the observed cancers near the plant.

In fact, such an expected death calculation has already been made by Devine *et al.* (1998), using traditional life table methods operating on the Rogers & Killough (1997) cell-level population estimates (which were in turn derived simply as proportional to the structure counts). However, these estimates were only for the total population in each cell; sex- and age group-specific counts were obtained by "breaking out" the totals into subcategories using a standard table (i.e., the *same* table in each cell, regardless of its true demographic makeup). In addition, the uncertainty associated with the cell-specific population estimates was quantified in a rather ad-hoc way through distributional assumptions on the true number of structures in each cell and the true occupancy rate per structure, with this uncertainty propagated through to the estimated numbers of lung cancer deaths. We instead adopt a Bayesian approach that allows not only formal prior-to-posterior updating of point estimates and standard errors, but full posterior distributions for the imputed population counts in the 160 target cells in the windrose. We are also able to incorporate other available prior and covariate information, and to benefit from the implicit spatial smoothing.

In what follows we create a "snapshot" of the population for the year 1990. This particular year was convenient for us because of the ready availability of census data and block group boundary locations from CD-ROMs and the World Wide Web. Multiyear modeling was beyond the scope of our project, since the Census Bureau typically changes the boundaries of its block groups with every census, and earlier boundaries are not available in computerized form. Nevertheless, it is possible to extend our methods to the temporal case.

1.3.2   Implementation Issues

We elucidate the notation of the previous section for our Section 1.3 dataset as follows. The responses Y are population counts, while the misaligned covariates X are structure counts; our model has no additional covariates W_i. In the context of Figure 1.3, the B_i are the block groups and the C_j are the windrose cells. The union of the census blocks contains the windrose ($S_C \subset S_B$), so only B-edge atoms are present. Finally, henceforth we emphasize the summing over atoms within a block group or cell by replacing the appropriate subscript with a dot (e.g., $X_{j\cdot} = \sum_{l=1}^{L_j} X_{jl}$).

The modeling ideas in Section 1.2 are implemented for the FMPC dataset with a Markov chain Monte Carlo approach, implemented with **Fortran** and **S-PLUS** routines. The basic structure is that of a Gibbs sampler, frequently employing Metropolis and Hastings substeps (see e.g. Carlin & Louis 2000, Sec. 5.4.4). We run five independent chains for all parameters, using trace plots, lag 1 sample autocorrelations, and Gelman & Rubin (1992) diagnostic statistics to assess convergence of the algorithm. As mentioned in Section 1.2 the $\{Y_i\} \rightarrow \{Y_{ik}\}$ step is predictive, which allows us to implement it after the model has been fitted. That is, the $\{Y_{ik}\}$ may be analytically integrated out of the full model specification (1.7), the sampler run on the resulting lower-dimensional parameter space, and the required $\{Y_{ik}\}$ samples drawn from the appropriate multinomial at the very end.

We faced several challenges in implementing our model on the FMPC dataset. The first was the amount of bookkeeping in the 'lookup' tables that allowed for atoms p to be referenced in any of three systems: p, (ik), and (jl). Introduction of a CAR prior on the $\{\omega_j\}$ and $\{\omega_i^*\}$ requires yet more lookup tables, as it is necessary to keep track of which regions are neighbors (the customary CAR model *adjacency matrix*). Making the assumption that adopting a CAR prior for the structure spatial effects will result in sufficient smoothing of the fitting maps, we simply placed an iid (not a CAR) prior on the $\{\mu_i\}$. Details concerning the number and types of necessary lookup tables are given in Mugglin (1999).

A second difficulty stemmed from the discrete multivariate (and sometimes zero-valued) nature of the multinomial distribution in the MCMC setting. This presented several obstacles, the first being the choice of candidate density for the Metropolis-Hastings rejection step. Initially we chose to propose structures in each cell as a multinomial value, allocating structures to atoms proportionally to area (i.e., equating the candidate and the prior distributions). This was convenient computationally, but did not produce acceptable convergence because in many cells the proposed values were not similar enough to the posteriors. A first-order Maclaurin approximation to provide multinomial candidates that were closer to the posterior also proved ineffective. We finally adopted a multinomial proposal based only on the present state of the chain. That is, if this state is $(X_{j1}, \ldots, X_{jL_j})$, we propose a multinomial value drawn from a $\mathrm{Mult}(X_{j\cdot}\,;\, q_{j1}, \ldots, q_{jL_j})$ where $q_{jl} = X_{jl}/X_{j\cdot\cdot}$. A further correction is required since whenever a present X_{jl} becomes 0, under this proposal it will be 0 in all subsequent iterations as well. Hence we modify q_{jl} to $q_{jl} = (X_{jl}+1)/(X_{j\cdot}+L_j)$. We emphasize that this modification affects *only* the Metropolis-Hastings proposal density, and not the model itself.

A somewhat different problem required a similar solution. We originally took the function h in (1.2) to be the identity function, producing the model $Y_{ik} \sim Po\,(e^{\mu_i}\,(X'_{ik}))$, which in turn implies $Y_{i\cdot} \sim Po\,(e^{\mu_i}(X'_{i\cdot}))$. Suppose however that $Y_{i\cdot} > 0$ for a particular block group i, but in some MCMC iteration no structures are allocated to any of the atoms of the block group. The result is a flawed probabilistic specification. To ensure $h > 0$ even when $z = 0$, we revised our model to $h(z\,;\,\theta_{ik}) = z + \theta_{ik}$ where $\theta_{ik} = \theta/(K_i|B_{ik}|)$ with $\theta > 0$, resulting in

$$Y_{ik} \sim Po\left(e^{\mu_i}\left(X'_{ik} + \frac{\theta}{K_i}\right)\right) . \tag{1.8}$$

This adjustment eliminates the possibility of a zero-valued Poisson parameter, but does allow for the possibility of a nonzero population count in a region where there are no structures observed. When conditioned on $Y_{i\cdot}$, we find $(Y_{i1}, \ldots, Y_{iK_i} \mid Y_{i\cdot}) \sim \mathrm{Mult}(Y_{i\cdot}\,;\, p_{i1}, \ldots, p_{iK_i})$, where

$$p_{ik} = \frac{X'_{ik} + \theta/K_i}{X'_{i\cdot} + \theta} \quad \text{and} \quad Y_{i\cdot} \sim Po\,(e^{\mu_i}(X'_{i\cdot} + \theta)) . \tag{1.9}$$

Our basic model then consists of (1.8) – (1.9) together with the following:

$$\mu_i \stackrel{iid}{\sim} N\left(\eta_\mu, 1/\tau_\mu\right), \quad X_{jl} \sim Po\left(e^{\omega_j}|C_{jl}|\right)$$
$$\Rightarrow X_{j\cdot} \sim Po\left(e^{\omega_j}|C_j|\right),$$
$$(X_{j1}, \ldots, X_{jL_j} \mid X_{j\cdot}) \sim \text{Mult}(X_{j\cdot}\, ; \, q_{j1}, \ldots, q_{jL_j}), \qquad (1.10)$$
$$\text{where } q_{jl} = |C_{jl}|/|C_j|,$$
$$X'_{iE} \sim Po\left(e^{\omega_i^*}|B_{iE}|\right), \text{ and } (\omega_j, \omega_i^*) \sim \text{CAR}(\lambda_\omega),$$

where X'_{iE} and ω_i^* refer to edge atom structure counts and log relative risk parameters, respectively. While θ could be estimated from the data, in our implementation we simply set $\theta = 1$; Section 1.6 discusses the impact of alternate selections.

A third implementation challenge we faced was the high autocorrelations in the sampled chains. We implemented a standard Metropolis-Hastings algorithm with five independent parallel chains. With the structures missing outside the windrose and highly constrained (by their cell totals) within the windrose, the algorithm appears to traverse the parameter space slowly: the median of the lag-1 autocorrelations over the 1006 parameters is 0.76; 8% of them are larger than 0.99. Wherever possible we tuned candidate densities so that proposed values were accepted at rates between 35% and 50%, as recommended by Gelman *et al.* (1996), as well as years of Metropolis "folklore." It should be noted, however, that the structures (proposed as multinomials in the windrose cells and as Poissons in the edge atoms) could not be tuned as with usual Gaussian or t proposals, because the variances could not be specified independently of the means. Fortunately in all cases we were still able to obtain a reasonable minimum level of candidate acceptance.

1.3.3   Data Analysis

We turn now to the particulars of the FMPC data analysis, examining three different models in the context of the misaligned data as described in Section 1.3. In the first case we take up the problem of total population interpolation, while in the second and third cases we consider age- and sex-specific population interpolation.

We begin by taking $\eta_\mu = 1.1$ and $\tau_\mu = 0.5$ in (1.10). The choice of mean value reflects the work of Rogers & Killough (1997), who found population per household (PPH) estimates for four of the seven townships in which the windrose lies. Their estimates ranged in value from 2.9 to 3.2, hence our choice of $\eta_\mu = 1.1 \approx \log(3)$. The value $\tau_\mu = 0.5$ is sufficiently small to make the prior for μ_i large enough to support all feasible values of μ_i (two prior standard deviations in either direction would enable PPH values of 0.18 to 50.8).

For $\omega = \{\omega_j, \omega_i^*\}$ we adopted a CAR prior and fixed $\lambda_\omega = 10$. We did not impose any centering of the elements of ω around 0, allowing them to determine their own mean level in the MCMC algorithm. Since most cells have four neighbors, the value $\lambda_\omega = 10$ translates into a conditional prior standard deviation for the ω's of $\sqrt{1/(10 \cdot 4)} = 0.158$, hence a marginal prior standard deviation of roughly $0.158/0.7 \approx 0.23$ (Bernardinelli *et al.* 1995). In any case, we found $\lambda_\omega < 10$ too vague to allow MCMC convergence. Typical posterior medians for the ω's ranged from 2.2 to 3.3 for the windrose ω_j's and from 3.3 to 4.5 for the edge ω_i^*s.

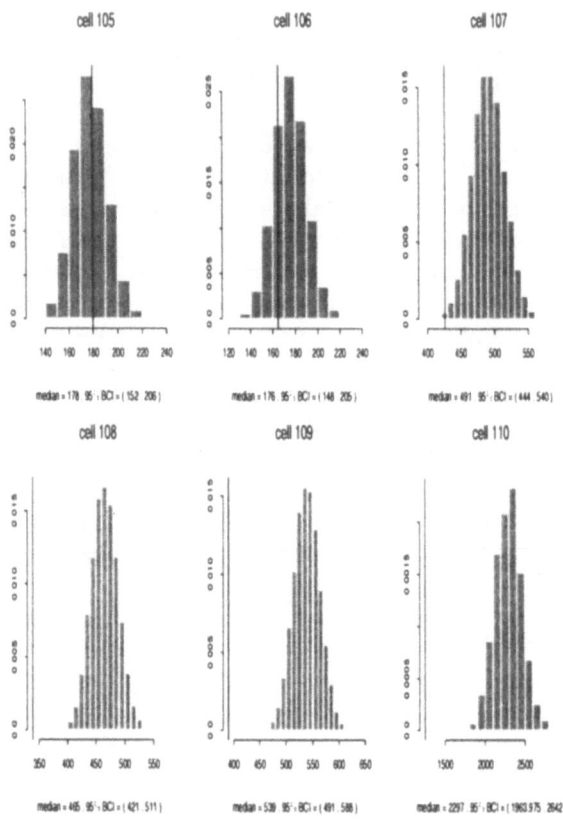

Fig. 1.4. Posterior distributions of populations in cells 105-110. Vertical bars represent estimates formed by multiplying structures per cell by a constant population per household (PPH) of 3.0.

Running 5 parallel sampling chains, acceptable convergence obtains for all parameters within 1500 iterations. We discarded this initial sample and then continued the chains for an additional 5000 iterations each, obtaining a final posterior sample of size 25,000. Population estimates per cell for cells 105 through 110 (again in the SE direction, from the middle to outer edge of the windrose) are indicated in Figure 1.4. Vertical bars here represent estimates

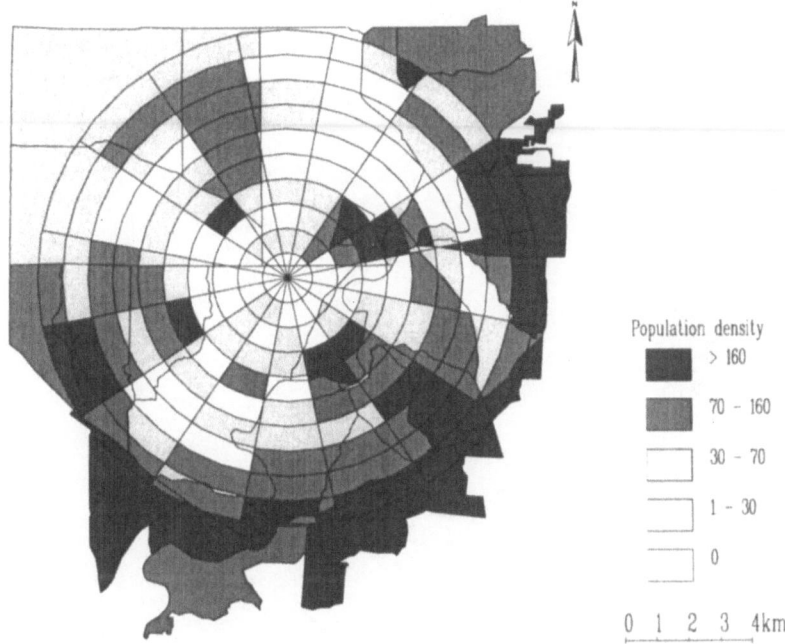

Fig. 1.5. Imputed population densities (persons/km^2) by atom for the FMPC region.

calculated by multiplying the number of structures in the cell by a fixed (map-wide) constant representing population per household (PPH), a method roughly equivalent to that employed by Rogers & Killough (1997), who as mentioned above actually used four different PPH values. Our reference lines use a constant value of 3 (the analogue of our prior mean). While cells 105 and 106 indicate good general agreement in these estimates, cells 107 through 110 display markedly different population estimates, where our estimates are substantially higher than the constant-PPH estimates. This is typical of cells toward the outer edge of the southeast portion of the windrose, since the suburbs of Cincinnati encroach on this region. We have population data only (no structures) in the southeastern edge atoms, so our model must estimate both the structures and the population in these regions. The resulting PPH is higher than a mapwide value of 3 (one would expect suburban PPH to be greater than rural PPH) and so the CAR model placed on the $\{\omega_j, \omega_i^*\}$ parameters induces a spatial similarity that can be observed in Figure 1.4.

We next implement the $\{Y_{i\cdot}\} \rightarrow \{Y_{ik}\}$ step. From the resulting $\{Y_{ik}\}$ come the $\{Y_j'\}$ cell totals by appropriate reaggregation. Figure 1.5 shows the population densities by atom ($Y_{ik}/|B_{ik}|$), calculated by taking the posterior medians of the population distributions for each atom and dividing by atom area in square kilometers. This figure clearly shows the encroachment by suburban

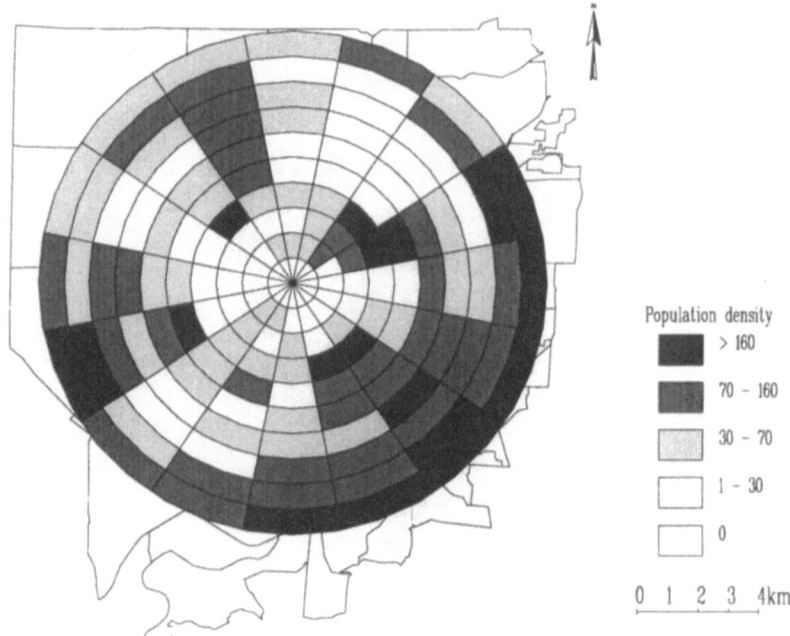

Fig. 1.6. Imputed population densities (persons/km^2) by cell for the FMPC windrose.

Cincinnati on the southeast side of our map, with some spatial smoothing between the edge cells and the outer windrose cells. Finally, Figure 1.6 shows population densities by cell ($Y'_j/|C_j|$), where the atom-level populations have been aggregated to cells before calculating densities. Posterior standard deviations, though not shown, are also available for each cell. While this figure, by definition, provides less detail than Figure 1.5, it provides information at the scale appropriate for combination with the exposure values of Killough *et al.* (1996). Moreover, the scale of aggregation is still fine enough to permit identification of the locations of Cincinnati suburban sprawl, as well as the communities of Ross (contained in cells ENE 4-5 and NE 4), Shandon (NW 4-5), New Haven (WSW 5-6), and New Baltimore (SSE 4-5).

1.4   Misaligned Point-Block Data Model Development

1.4.1   Model Assumptions and Analytic Goals

The setting we work with is to assume that we are observing a continuous variable and that underlying all observations of this variable is a spatial process. We denote this process by $Y(s)$ for locations $s \in D$, a region of interest. In our applications $D \subset R^2$ but our development works in arbitrary dimensions. A realization of the process is a surface over D. For point-referenced data the realization is observed at a finite set of sites, say $s_i, i = 1, 2, \ldots, I$. For block data we assume the observations arise as block averages. That is, for a block $B \subset D$,

$$Y(B) = |B|^{-1} \int_B Y(s)ds \qquad (1.11)$$

where $|B|$ denotes the area of B (see e.g. Cressie 1993). The integration in (1.11) is an average of random variables, hence a random or stochastic integral. Thus, the assumption of an underlying spatial process is only appropriate for block data that can be sensibly viewed as an averaging over point data; examples of this would include rainfall, pollutant level, temperature and elevation. It would be inappropriate for, say, population, since there is no "population" at a particular point. It would also be inappropriate for most proportions of interest. For instance, if $Y(B)$ is the proportion of college-educated persons in B, then $Y(B)$ is continuous but $Y(s)$, assuming a single individual at a point, is binary.

In general, we envision four possibilities. First, starting with point data $Y(s_1), \ldots, Y(s_I)$, we seek to predict at new locations, i.e., to infer about $Y(s_1'), \ldots, Y(s_K')$ (points to points). Second, starting with point data, we seek to predict at blocks, i.e., to infer about $Y(B_1), \ldots, Y(B_K)$ (points to blocks). Third, starting with block data $Y(B_1), \ldots, Y(B_I)$, we seek to predict at a set of locations, i.e., to infer about $Y(s_1'), \ldots, Y(s_K')$ (blocks to points). Finally, starting with block data, we seek to predict at new blocks, i.e., to infer about $Y(B_1'), \ldots, Y(B_K')$ (blocks to blocks).

A new wrinkle we add to the change of support problem here is application in the context of spatio-temporal data. In our Atlanta dataset, the number of monitoring stations is small but the amount of data collected over time is substantial. In this case, under suitable modeling assumptions, we may not only learn about the temporal nature of the data but also enhance our understanding of the spatial process. Moreover, the additional computational burden to analyze the much larger dataset within the Bayesian framework turns out still to be manageable.

More specifically, we illustrate with point-referenced data, assuming observations $Y(s_i, t_j)$ at locations s_i, $i = 1, \ldots, I$, and at times t_j, $j = 1, \ldots, J$. Thus, we assume all locations are monitored at the same set of times. In the sequel, we adopt an equally spaced time scale, so that we are modeling a time series of spatial processes. In our development we assume that all IJ points have been observed; randomly missing values create no additional conceptual difficulty, but add computational burden. As a result, the change of support problem is only in space. At a new location s' we seek to predict the entire vector $(Y(s', t_1), \ldots, Y(s', t_J))$. Similarly, at a block B we seek to predict the vector of block averages $(Y(B, t_1), \ldots, Y(B, t_J))$.

1.4.2   Methodology for the Static Spatial Case

We start with a stationary Gaussian process specification for $Y(s)$ having mean function $\mu(s; \beta)$ and covariance function $c(s - t; \theta)$. Here μ is a trend surface with coefficient vector β while θ denotes the parameters associated with the stationary covariance function. Beginning with point data observed at sites $s_1, ..., s_I$, let $Y_s^T = (Y(s_1), ..., Y(s_I))$. Then

$$Y_s \mid \beta, \theta \sim N(\mu_s(\beta), H_s(\theta)) \tag{1.12}$$

where $\mu_s(\beta)_i = \mu(s_i; \beta)$ and $(H_s(\theta))_{ii'} = c(s_i - s_{i'}; \theta)$.

Given a prior on β and θ, models such as (1.12) are straightforwardly fit using simulation methods either through importance sampling (e.g., Ecker & Gelfand, 1997) or Gibbs sampling (e.g., Karson *et al.*, 1999). Regardless, we can assume the existence of posterior samples (β_g^*, θ_g^*), $g = 1, ..., G$ from $f(\beta, \theta \mid Y_s)$. The required inversion of $H_s(\theta)$ in order to evaluate the likelihood can be problematic when I is large.

Then for prediction at a set of new locations, $Y_{s'}^T = (Y(s'_1), ..., Y(s'_K))$, we require only the predictive distribution

$$f(Y_{s'} \mid Y_s) = \int f(Y_{s'} \mid Y_s, \beta, \theta) f(\beta, \theta \mid Y_s) d\beta d\theta . \tag{1.13}$$

By drawing $Y_{s',g}^* \sim f(Y_{s'} \mid Y_s, \beta_g^*, \theta_g^*)$ we obtain a sample from (1.13) which provides any desired inference about $Y_{s'}$ and its components.

Under a Gaussian process,

$$f\left(\begin{pmatrix} Y_s \\ Y_{s'} \end{pmatrix} \middle| \beta, \theta\right) = N\left(\begin{pmatrix} \mu_s(\beta) \\ \mu_{s'}(\beta) \end{pmatrix}, \begin{pmatrix} H_s(\theta) & H_{s,s'}(\theta) \\ H_{s,s'}^T(\theta) & H_{s'}(\theta) \end{pmatrix}\right) \tag{1.14}$$

with entries defined as in (1.12). Hence, $Y_{s'} \mid Y_s, \beta, \theta$ is distributed as

$$N\left(\mu_{s'}(\beta) + H_{s,s'}^T(\theta) H_s^{-1}(\theta)(Y_s - \mu_s(\beta)), \atop H_{s'}(\theta) - H_{s,s'}^T(\theta) H_s^{-1}(\theta) H_{s,s'}(\theta)\right) . \tag{1.15}$$

Sampling from (1.15) requires the inversion of $H_s(\theta_g^*)$, which will already have been done in sampling θ_g^*, and then the square root of the $K \times K$ covariance matrix in (1.15).

Turning to prediction for blocks, say at $Y_B^T = (Y(B_1), ..., Y(B_K))$ we again require the predictive distribution, which is now

$$f(Y_B \mid Y_s) = \int f(Y_B \mid Y_s; \beta, \theta) f(\beta, \theta \mid Y_s) d\beta d\theta . \tag{1.16}$$

Under a Gaussian process, we now have

$$f\left(\begin{pmatrix} Y_s \\ Y_B \end{pmatrix} \middle| \beta, \theta\right) = N\left(\begin{pmatrix} \mu_s(\beta) \\ \mu_B(\beta) \end{pmatrix}, \begin{pmatrix} H_s(\theta) & H_{s,B}(\theta) \\ H_{s,B}^T(\theta) & H_B(\theta) \end{pmatrix}\right) , \tag{1.17}$$

where

$$(\mu_B(\beta))_k = E(Y(B_k) \mid \beta) = |B_k|^{-1} \int_{B_k} \mu(s; \beta)ds \ ,$$

$$(H_B(\theta))_{kk'} = |B_k|^{-1}|B_{k'}|^{-1} \int_{B_k} \int_{B_{k'}} c(s - s'; \theta)ds'ds \ ,$$

and $$(H_{s,B}(\theta))_{ik} = |B_k|^{-1} \int_{B_k} c(s_i - s'; \theta)ds' \ .$$

Analogously to (1.15), $\boldsymbol{Y}_B | \boldsymbol{Y}_s, \beta, \theta$ is distributed as

$$N\left(\mu_B(\beta) + H_{s,B}^T(\theta)H_s^{-1}(\theta)(\boldsymbol{Y}_s - \mu_s(\beta)) \ , \atop H_B(\theta) - H_{s,B}^T(\theta)H_s^{-1}(\theta)H_{s,B}(\theta)\right) \ . \tag{1.18}$$

The major difference between (1.15) and (1.18) is that in (1.15), given (β_g^*, θ_g^*), numerical values for all of the entries in $\mu_{s'}(\beta)$, $H_{s'}(\theta)$, and $H_{s,s'}(\theta)$ are immediately obtained. In (1.18) every analogous entry requires an integration as above. Anticipating irregularly shaped B_k's, Riemann approximation to integrate over these regions may be awkward. Instead, noting that each such integration is an expectation with respect to a uniform distribution, we propose Monte Carlo integration. In particular, for each B_k we propose to draw a set of locations $s_{k,\ell}$, $\ell = 1, 2, ..., L_k$, distributed independently and uniformly over B_k. Here L_k can vary with k to allow for very unequal $|B_k|$. Hence, we replace $(\mu_B(\beta))_k$, $(H_B(\theta))_{kk'}$, and $(H_{s,B}(\theta))_{ik}$ with

$$(\widehat{\mu}_B(\beta))_k = L_k^{-1} \sum_\ell \mu(s_{k,\ell}; \beta) \ , \atop (\widehat{H}_B(\theta))_{kk'} = L_k^{-1}L_{k'}^{-1} \sum_\ell \sum_{\ell'} c(s_{k\ell} - s_{k'\ell'}; \theta) \atop \text{and } (\widehat{H}_{s,B}(\theta))_{ik} = L_k^{-1} \sum_\ell c(s_i - s_{k\ell}; \theta) \ . \tag{1.19}$$

In our notation, the 'hat' denotes a Monte Carlo integration which can be made arbitrarily accurate and has nothing to do with the data $\boldsymbol{Y}_s$. Note also that the same set of $s_{k\ell}$'s can be used for each integration and with each (β_g^*, θ_g^*); we need only obtain this set once. In obvious notation we replace (1.17) with the $(I + K)$-dimensional multivariate normal distribution $\widehat{f}\left((\boldsymbol{Y}_s, \boldsymbol{Y}_B)^T \mid \beta, \theta\right)$.

It is useful to note that, if we define $\widehat{Y}(B_k) = L_k^{-1} \sum_\ell Y(s_{k\ell})$, then $\widehat{Y}(B_k)$ is a Monte Carlo integration for $Y(B_k)$ as given in (1.11). Then, with an obvious definition for $\widehat{\boldsymbol{Y}}_B$, it is apparent that

$$\widehat{f}((\boldsymbol{Y}_s, \boldsymbol{Y}_B)^T \mid \beta, \theta) = f\left((\boldsymbol{Y}_s, \widehat{\boldsymbol{Y}}_B)^T \mid \beta, \theta\right) \tag{1.20}$$

where (1.20) means that the approximate joint distribution of $(\boldsymbol{Y}_s, \boldsymbol{Y}_B)$ is the exact joint distribution of $(\boldsymbol{Y}_s, \widehat{\boldsymbol{Y}}_B)$. In practice, we will work with $\widehat{f}$, converting to $\widehat{f}(\boldsymbol{Y}_B \mid \boldsymbol{Y}_s, \beta, \theta)$ to sample $\boldsymbol{Y}_B$ rather than sampling the $\widehat{Y}(B_k)$'s through the $Y(s_{k\ell})$'s. But, evidently, we are sampling $\widehat{\boldsymbol{Y}}_B$ rather than $\boldsymbol{Y}_B$.

Finally, suppose we start with block data, $\boldsymbol{Y}_B^T = (Y(B_1), \ldots, Y(B_I))$. Then, analogous to (1.12), the likelihood is well defined, i.e.,

$$f(\boldsymbol{Y}_B \mid \beta, \theta) = N(\mu_B(\beta), H_B(\theta)). \qquad (1.21)$$

Hence, given a prior on β and θ, the Bayesian model is completely specified. As above, evaluation of the likelihood requires integrations. So, we replace (1.21) with

$$\widehat{f}(\boldsymbol{Y}_B \mid \beta, \theta) = N(\widehat{\mu}_B(\beta), \widehat{H}_B(\theta)). \qquad (1.22)$$

Simulation-based fitting is now straightforward, as below (1.12), albeit somewhat more time-consuming due to the need to calculate $\widehat{\mu}_B(\beta)$ and $\widehat{H}_B(\theta)$. To predict for $\boldsymbol{Y}_{s'}$ we require $f(\boldsymbol{Y}_{s'} \mid \boldsymbol{Y}_B)$, hence $f(\boldsymbol{Y}_B, \boldsymbol{Y}_{s'} \mid \beta, \theta)$, which has been given in (1.17). Using (1.20) we now obtain $\widehat{f}(\boldsymbol{Y}_{s'} \mid \boldsymbol{Y}_B, \beta, \theta)$ to sample $\boldsymbol{Y}_{s'}$. Note that $\widehat{f}$ is used in (1.22) to obtain the posterior samples and again to obtain the predictive samples. Equivalently, the foregoing discussion shows that we can replace $\boldsymbol{Y}_B$ with $\widehat{\boldsymbol{Y}}_B$ throughout. To predict for new blocks $B_1', \ldots, B_K'$, let $\boldsymbol{Y}_{B'}^T = (Y(B_1'), \ldots, Y(B_K'))$. Now we require $f(\boldsymbol{Y}_{B'} \mid \boldsymbol{Y}_B)$, which in turn requires $f(\boldsymbol{Y}_B, \boldsymbol{Y}_{B'} \mid \beta, \theta)$. The approximate distribution $\widehat{f}(\boldsymbol{Y}_B, \boldsymbol{Y}_{B'} \mid \beta, \theta)$ employs Monte Carlo integrations over the B_k''s as well as the B_i's, and yields $\widehat{f}(\boldsymbol{Y}_{B'} \mid \boldsymbol{Y}_B, \beta, \theta)$ to sample $\boldsymbol{Y}_{B'}$. Again $\widehat{f}$ is used to obtain both the posterior and predictive samples.

Note that, in all four prediction cases, we can confine ourselves to an $(I + K)$-dimensional multivariate normal. Moreover, we have only an $I \times I$ matrix to invert repeatedly in the model fitting, and a $K \times K$ matrix whose square root is required for the predictive sampling.

1.4.3   Methodology for Spatio-Temporal Data

Following Subsection 1.4.2, we start with point-referenced data, $Y(\boldsymbol{s}_i, t_j)$, $i = 1, 2, \ldots, I$, $j = 1, 2, \ldots, J$ and specify a stationary Gaussian multivariate process for $\boldsymbol{Y}(\boldsymbol{s})^T = (Y(\boldsymbol{s}, t_1), \ldots, Y(\boldsymbol{s}, t_j))$. In particular, we set $E(Y(\boldsymbol{s}, t) \mid \beta) = \mu(\boldsymbol{s}, t; \beta)$, a temporal trend surface model (i.e., a polynomial in $\boldsymbol{s}$ and t). As for the cross-covariance structure, i.e., the matrix of covariances between $\boldsymbol{Y}(\boldsymbol{s}_i)$ and $\boldsymbol{Y}(\boldsymbol{s}_{i'})$, we assume the (j, j') entry,

$$cov(Y(\boldsymbol{s}_i, t_j), Y(\boldsymbol{s}_{i'}, t_{j'})) = \sigma^2 c^{(1)}(\boldsymbol{s}_i - \boldsymbol{s}_{i'}; \phi) \cdot c^{(2)}(t_j - t_{j'}; \rho), \qquad (1.23)$$

where $c^{(1)}$ is a valid two-dimensional correlation function and $c^{(2)}$ is a valid one-dimensional correlation function. The multiplicative form conveniently separates space and time in the calculations below, enabling feasible computation for the change of support problem. In view of the discreteness of the time scale, (1.23) suggests that we may view the entire specification as a time series of spatial processes. Spatial association at a fixed time point is captured through $c^{(1)}$; decay in such association over time is captured by

$c^{(2)}$. Forms such as (1.23) have a history in spatio-temporal modeling; see e.g. Mardia & Goodall (1993).

Collecting the observed data into a vector $\boldsymbol{Y}_s^T = (\boldsymbol{Y}^T(\boldsymbol{s}_1), \ldots, \boldsymbol{Y}^T(\boldsymbol{s}_I))$ we have $\boldsymbol{Y}_s$ distributed as an IJ-dimensional multivariate normal with, in obvious notation, mean vector $\mu_s(\beta)$ and covariance matrix

$$\Sigma_{\boldsymbol{Y}_s}(\sigma^2, \phi, \rho) = \sigma^2 H_s(\phi) \otimes H_t(\rho) , \tag{1.24}$$

where "$\otimes$" denotes the Kronecker product. In (1.24), $H_s(\phi)$ is $I \times I$ with $(H_s(\phi))_{ii'} = c^{(1)}(\boldsymbol{s}_i - \boldsymbol{s}_i'; \theta)$, and $H_t(\rho)$ is $J \times J$ with $(H_t(\rho))_{jj'} = c^{(2)}(t_j - t_{j'}; \rho)$.

Given a prior for β, σ^2, ϕ, and ρ, the Bayesian model is completely specified. Simulation-based model fitting can be carried out similarly to the static spatial case by noting the following. The log likelihood arising from $\boldsymbol{Y}_s$ is

$$-\tfrac{1}{2} \log \left| \sigma^2 H_s(\phi) \otimes H_t(\rho) \right|$$
$$-\tfrac{1}{2\sigma^2} (\boldsymbol{Y}_s - \mu_s(\beta))^T (H_s(\phi) \otimes H_t(\rho))^{-1} (\boldsymbol{Y}_s - \mu_s(\beta)) .$$

But $\left| \sigma^2 H_s(\phi) \otimes H_t(\rho) \right| = (\sigma^2)^{IJ} |H_s(\phi)|^J |H_t(\rho)|^I$, and $(H_s(\phi) \otimes H_t(\rho))^{-1} = H_s^{-1}(\phi) \otimes H_t^{-1}(\rho)$. In other words, even though (1.24) is $IJ \times IJ$, we need only the determinant and inverse for an $I \times I$ and a $J \times J$ matrix, so that Gibbs sampling is tractable.

With regard to prediction, first consider new locations $\boldsymbol{s}_1', \ldots, \boldsymbol{s}_k'$ with interest in inference for $Y(\boldsymbol{s}_k', t_j)$. As with the observed data, we collect the $Y(\boldsymbol{s}_k', t_j)$ into vectors $\boldsymbol{Y}(\boldsymbol{s}_k')$, and the $\boldsymbol{Y}(\boldsymbol{s}_k')$ into a single $KJ \times 1$ vector $\boldsymbol{Y}_{s'}$. Even though we may not necessarily be interested in every component of $\boldsymbol{Y}_{s'}$, the simplifying forms which follow below suggest that, with regard to programming, it may be easiest to simulate draws from the entire predictive distribution $f(\boldsymbol{Y}_{s'} \mid \boldsymbol{Y}_s)$ and then retain only the desired components.

Given posterior samples $(\beta_g^*, \sigma_g^{2*}, \phi_g^*, \rho_g^*)$, $g = 1, \ldots, G$, since $f(\boldsymbol{Y}_{s'} | \boldsymbol{Y}_s)$ is analogous to (1.13) we may draw $\boldsymbol{Y}_{s',g}^*$ from $f(\boldsymbol{Y}_{s'} \mid \boldsymbol{Y}_s, \beta_g^*, \sigma_g^{2*}, \phi_g^*, \rho_g^*)$. Analogous to (1.14), $f\left((\boldsymbol{Y}_s, \boldsymbol{Y}_{s'})^T | \beta, \sigma^2, \phi, \rho\right)$ is

$$N\left(\begin{pmatrix} \mu_s(\beta) \\ \mu_{s'}(\beta) \end{pmatrix} , \sigma^2 \begin{pmatrix} H_s(\phi) \otimes H_t(\rho) & H_{s,s'}(\phi) \otimes H_t(\rho) \\ H_{s,s'}^T(\phi) \otimes H_t(\rho) & H_{s'}(\phi) \otimes H_t(\rho) \end{pmatrix} \right) , \tag{1.25}$$

with obvious definitions for $H_{s'}(\phi)$ and $H_{s,s'}(\phi)$. Thus $\boldsymbol{Y}_{s'} \mid \boldsymbol{Y}_s, \beta, \sigma^2, \phi, \rho$ is also normally distributed, with mean

$$\mu_{s'}(\beta) + (H_{s,s'}^T(\phi) \otimes H_t(\phi))(H_s(\phi) \otimes H_t(\rho))^{-1}(Y_s - \mu_s(\beta))$$
$$= \mu_{s'}(\beta) + (H_{s,s'}^T(\phi) H_s^{-1}(\phi) \otimes I_{J \times J})(\boldsymbol{Y}_s - \mu_s(\beta)) , \tag{1.26}$$

and variance

$$H_{s'}(\phi) \otimes H_t(\rho) - (H_{s,s'}^T \otimes H_t(\rho))(H_s(\phi) \otimes H_t(\rho))^{-1}(H_{s,s'}(\phi) \otimes H_t(\rho))$$
$$= (H_{s'}(\phi) - H_{s,s'}^T(\phi) H_s^{-1}(\phi) H_{s,s}(\phi)) \otimes H_t(\rho) , \tag{1.27}$$

using standard properties of Kronecker products. In (1.26), time disappears apart from $\mu_{s'}(\beta)$, while in (1.27), time "factors out" of the conditioning. Sampling from this normal distribution does require the inverse square root of the conditional covariance matrix, but conveniently, this is

$$(H_{s'}(\phi) - H_{s,s'}^T(\phi)H_s^{-1}(\phi)H_{s,s'}(\phi))^{-\frac{1}{2}} \otimes H_t^{-\frac{1}{2}}(\rho) \, ,$$

so the only work required beyond that in (1.15) is obtaining $H_t^{-\frac{1}{2}}(\rho)$, since $H_t^{-1}(\rho)$ will already have been obtained in evaluating the likelihood.

For blocks $B_1, \ldots, B_K$, we set $\boldsymbol{Y}^T(B_k) = (Y(B_k, t_1), \ldots, Y(B_k, t_J))$ and then set $\boldsymbol{Y}_B^T = (\boldsymbol{Y}^T(B_1), \ldots, \boldsymbol{Y}^T(B_K))$. Analogous to (1.16) we seek to sample $f(\boldsymbol{Y}_B \mid \boldsymbol{Y}_s)$, so we require $f(\boldsymbol{Y}_B \mid \boldsymbol{Y}_s, \beta, \sigma^2, \phi, \rho)$. Analogous to (1.25), this is obtainable from $f\left((\boldsymbol{Y}_s, \boldsymbol{Y}_B)^T \mid \beta, \sigma^2, \phi, \rho\right)$, which is

$$N\left(\begin{pmatrix} \mu_s(\beta) \\ \mu_B(\beta) \end{pmatrix} , \ \sigma^2 \begin{pmatrix} H_s(\phi) \otimes H_t(\rho) & H_{s,B}(\phi) \otimes H_t(\rho) \\ H_{s,B}^T(\phi) \otimes H_t(\rho) & H_B(\phi) \otimes H_t(\rho) \end{pmatrix}\right) ,$$

with $\mu_B(\beta)$, $H_B(\phi)$, and $H_{s,B}(\phi)$ as in Subsection 1.4.2. Hence the distribution $f(\boldsymbol{Y}_B \mid \boldsymbol{Y}_s, \beta, \sigma^2, \phi, \rho)$ is again normal with mean and variance as given in (1.26) and (1.27), but with $\mu_B(\beta)$ replacing $\mu_{s'}(\beta)$, $H_B(\phi)$ replacing $H_{s'}(\phi)$, and $H_{s,B}(\phi)$ replacing $H_{s,s'}(\phi)$. Using the same Monte Carlo integrations as proposed in Subsection 1.4.2 leads to sampling the resultant $\hat{f}(\boldsymbol{Y}_B \mid \boldsymbol{Y}_s, \beta, \sigma^2, \phi, \rho)$, and the same technical justification applies.

If we started with block data, $Y(B_i, t_j)$, then following (1.21) and (1.24),

$$f(\boldsymbol{Y}_B \mid \beta, \sigma^2, \phi, \rho) = N(\mu_B(\beta) \, , \ \sigma^2(H_B(\phi) \otimes H_t(\rho)) \, . \tag{1.28}$$

Given (1.28), the path for prediction at new points or at new blocks is clear, following the above and the end of Subsection 1.4.2; we omit the details.

1.5   Example: Ozone Exposure by Zip Code in Atlanta
1.5.1   Motivating Dataset

A solution to the change of support problem is required in many health science applications, particularly spatial and environmental epidemiology. To illustrate, we consider a dataset of ozone levels in the Atlanta, GA metropolitan area, as reported by Tolbert *et al.* (2000). Ozone measures are available at between 8 and 10 fixed monitoring sites during the 92 summer days (June 1 through August 31) of 1995. Figure 1.7 shows the 1-hour daily maximum ozone measures at the 10 monitoring sites on July 15, 1995, along with the boundaries of the 162 zip codes in the Atlanta metropolitan area. Here we might be interested in predicting the ozone level at different points on the map (say, the two points marked A and B, which lie on opposite sides of a single city zip), or the average ozone level over a particular zip (say, one of the 36 zips falling within the city of Atlanta, the collection of which are

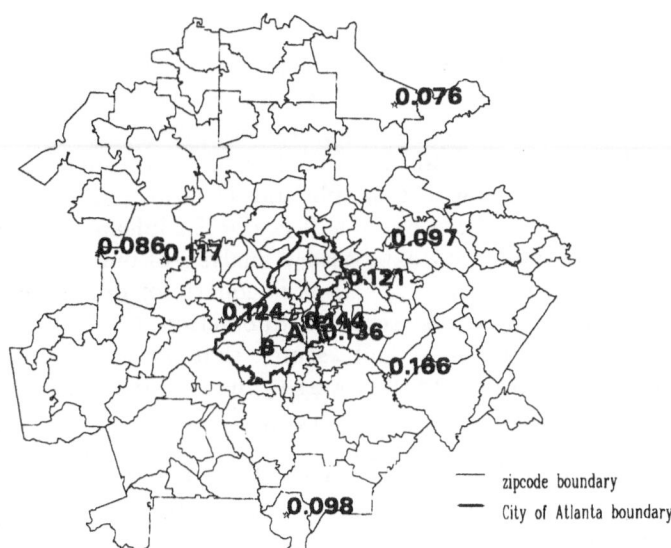

Fig. 1.7. Zip code boundaries in the Atlanta metropolitan area and ozone levels at the 10 monitoring sites for July 15, 1995.

encircled by the dark boundary on the map). The latter problem is of special interest, since in this case relevant health outcome data are available only at the zip level. In particular, for each day and zip, we have the number of pediatric ER visits for asthma, as well as the total number of pediatric ER visits. Thus an investigation of the relationship between ozone exposure and pediatric asthma cannot be undertaken until the mismatch in the support of the two variables is resolved. Situations like this are relatively common, since personal privacy concerns often limit statisticians' access to health outcome data other than at the block average level.

A previous study of this dataset by Carlin *et al.* (1999) realigned the point-level ozone measures to the zip level by using an `ARC/INFO` universal kriging procedure to fit a smooth ozone exposure surface, and subsequently took the kriged value at each zip centroid as the ozone value for that zip. But this approach uses a single centroid value to represent the ozone level in the entire zip, and fails to properly capture variability and spatial association by treating these kriged estimates as observed values.

1.5.2   Static Spatial Case

In this subsection, we use the approach of Subsection 1.4.2 to perform point-point and point-block inference for the Atlanta ozone data pictured in Figure 1.7. Recall that the target points are those marked A and B on the map, while the target blocks are the 36 Atlanta city zips. The differing block sizes suggest use of a different L_k for each k in equation (1.19). Conveniently, our

GIS, ARC/INFO, can generate random points over the whole study area, and then allocate them to each zip. Thus L_k is proportional to the area of the zip, $|B_k|$. Our procedure produced 3743 randomly chosen locations distributed over the 36 city zips, for an average L_k of nearly 104.

Suppose that log-ozone exposure $Y(s)$ follows a second-order stationary spatial Gaussian process, using the simple spatial covariance function $c(s_i - s_{i'}; \theta) = \sigma^2 e^{-\phi \| s_i - s_{i'} \|}$, where $\| s_i - s_{i'} \|$ is the Euclidean distance between sites s_i and $s_{i'}$. A preliminary exploratory analysis of our dataset suggested a constant mean function $\mu(s_i; \beta) = \mu$ is adequate for our dataset; c.f. Figure 1.9 and the associated discussion in Subsection 1.5.3 below. We place the customary flat prior on μ, and assume that $\sigma^2 \sim IG(a, b)$ and $\phi \sim G(c, d)$ where IG and G denote the inverse gamma and gamma distributions with probability density functions $p(\sigma^2 | a, b) = \frac{e^{-1/(b\sigma^2)}}{\Gamma(a) b^a (\sigma^2)^{a+1}}$ and $p(\phi | c, d) = \frac{\phi^{c-1} e^{-\phi/d}}{\Gamma(c) d^c}$, respectively. We chose $a = 3$, $b = 0.5$, $c = 0.03$ and $d = 100$, corresponding to fairly vague priors. We then fit this three-parameter model using an MCMC implementation, which ran 3 parallel sampling chains for 1000 iterations each, sampling μ and σ^2 via Gibbs steps and ϕ through Metropolis-Hastings steps with a $G(3, 1)$ candidate density. Convergence of the sampling chains was virtually immediate. We obtained the following posterior medians and 95% equal-tail credible intervals for the three parameters: for μ, 0.111 and (0.072, 0.167); for σ^2, 1.37 and (1.18, 2.11); and for ϕ, 1.62 and (0.28, 4.13).

Figure 1.8 maps summaries of the posterior samples for the 36 target blocks (city zips) and the 2 target points (A and B); specifically, the posterior medians, $q_{.50}$, upper and lower 0.025 points, $q_{.975}$ and $q_{.025}$, and the lengths of the 95% equal-tail credible intervals, $q_{.975} - q_{.025}$. The zip-level medians show a clear spatial pattern, with the highest predicted block averages occurring in the southeastern part of the city near the two high observed readings (0.144 and 0.136), and the lower predictions in the north apparently the result of smoothing toward the low observed value in this direction (0.076). The interval lengths reflect spatial variability, with lower values occurring in larger areas (which require more averaging) or in areas nearer to observed monitoring stations (e.g., those near the southeastern, northeastern, and western city boundaries). Finally, note that our approach allows sensibly differing predicted medians for points A and B, with A being higher due to the slope of the fitted surface. Previous centroid-based analyses (like that of Carlin *et al.*, 1999) would instead implausibly impute the same fitted value to both points, since both lie within the same zip.

1.5.3   Spatio-Temporal Case

To illustrate the method of Subsection 1.4.3, we use a spatio-temporal version of the Atlanta ozone dataset. As mentioned in Subsection 1.5.1, we actually

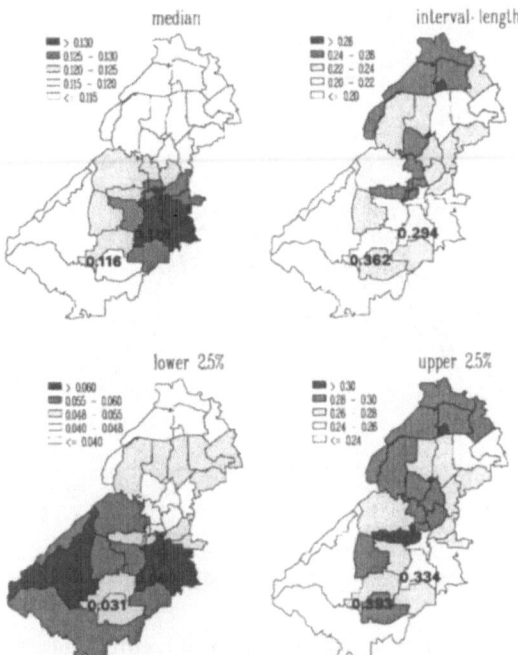

Fig. 1.8. Posterior point-point and point-block summaries, static spatial model, Atlanta ozone data for July 15, 1995.

have ozone measurements at the 10 fixed monitoring stations shown in Figure 1.7 over the 92 summer days in 1995. Figure 1.9 shows the daily one-hour maximum ozone reading for the sites during July of this year. There are several sharp peaks, but little evidence of a weekly (7-day) period in the data. The mean structure appears reasonably constant in space, with the ordering of the site measurements changing dramatically for different days. Moreover, with only 10 "design points" in the metro area, any spatial trend surface we fit would be quite speculative over much of the study region (e.g., the northwest and southwest metro; see Figure 1.7). The temporal evolution of the series is not inconsistent with a constant mean autoregressive error model; indeed, the lag 1 sample autocorrelation varies between 0.27 and 0.73 over the 10 sites, strongly suggesting the need for a model accounting for both spatial and temporal correlations.

We thus fit our spatio-temporal model with mean $\mu(\boldsymbol{s}, t; \beta) = \mu$, but with spatial and temporal correlation functions $c^{(1)}(\boldsymbol{s}_i - \boldsymbol{s}_{i'}; \phi) = e^{-\phi\|\boldsymbol{s}_i - \boldsymbol{s}_{i'}\|}$ and $c^{(2)}(t_j - t_{j'}; \rho) = \rho^{|j-j'|}/(1 - \rho^2)$. Hence our model has four parameters: we use a flat prior for μ, an $IG(3, 0.5)$ prior for σ^2, a $G(0.003, 100)$ prior for ϕ, and a $U(0, 1)$ prior for ρ (thus eliminating the implausible possibility of *negative* autocorrelation in our data, but favoring no positive value over any other). To facilitate our Gibbs-Metropolis approach, we transform to

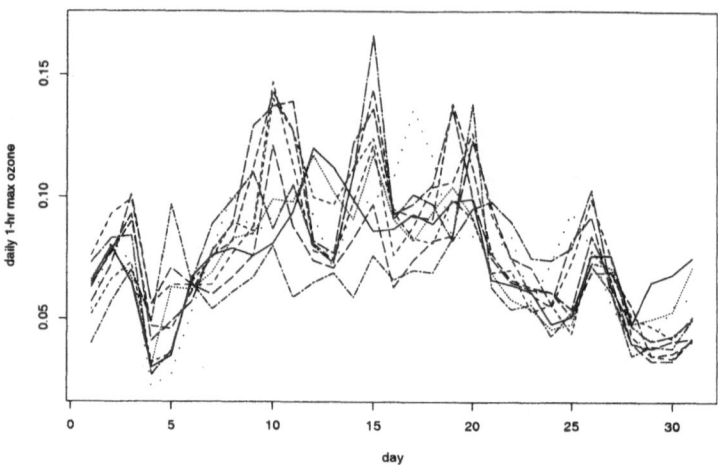

Fig. 1.9. One-hour maximum ozone by day, July 1995, 10 Atlanta monitoring sites.

Table 1.1. Posterior medians and 95% equal-tail credible intervals for ozone levels at two points, and for average ozone levels over three blocks (zip codes), purely spatial model versus spatio-temporal model, Atlanta ozone data for July 15, 1995.

	spatial only		spatio-temporal	
	point	95% interval	point	95% interval
Point A	.125	(.040, .334)	.139	(.111, .169)
Point B	.116	(.031, .393)	.131	(.098, .169)
Zip 30317 (east-central)	.130	(.055, .270)	.138	(.121, .155)
Zip 30344 (south-central)	.123	(.055, .270)	.135	(.112, .161)
Zip 30350 (north)	.112	(.040, .283)	.109	(.084, .140)

$\theta = \log \phi$ and $\lambda = \log(\rho/(1 - \rho))$, and subsequently use Gaussian proposals on these transformed parameters.

Running 3 parallel chains of 10,000 iterations each, sample traces (not shown) again indicate virtually immediate convergence of our algorithm. Posterior medians and 95% equal-tail credible intervals for the four parameters are as follows: for μ, 0.068 and (0.057, 0.080); for σ^2, 0.11 and (0.08, 0.17); for ϕ, 0.06 and (0.03, 0.08); and for ρ, 0.42 and (0.31, 0.52). The rather large value of ρ confirms the strong temporal autocorrelation suspected in the daily ozone readings.

Comparison of the posteriors for σ^2 and ϕ with those obtained for the static spatial model in Subsection 1.5.2 is not apt, since these parameters have different meanings in the two models. Instead, we make this comparison in the context of point-point and point-block prediction. Table 1.1 provides posterior predictive summaries for the ozone concentrations for July 15, 1995

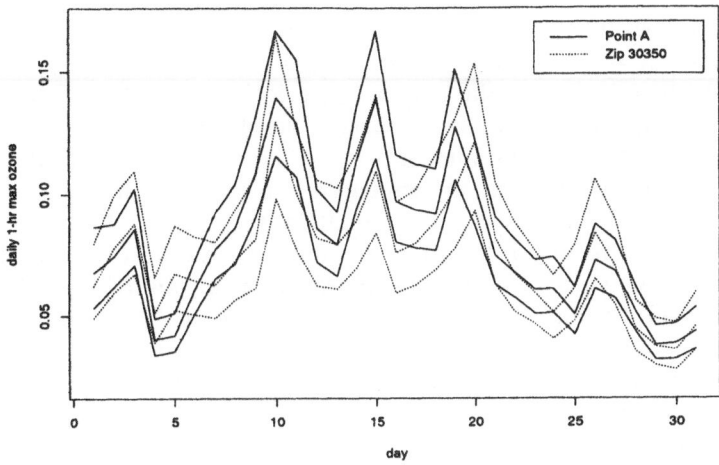

Fig. 1.10. Posterior medians and upper and lower .025 quantiles for the predicted one-hour maximum ozone concentration by day, July 1995; solid lines, point A; dotted lines, block average over zip 30350 (northernmost Atlanta city zip).

at points A and B (see Figure 1.7), as well as for the block averages over 3 selected Atlanta city zips: 30317, an east-central city zip very near to two monitoring sites; 30344, the south-central zip containing the points A and B; and 30350, the northernmost city zip. Results are shown for both the spatio-temporal model of this subsection and for the static spatial model previously fit in Subsection 1.5.2. Note that all the posterior medians are a bit higher under the spatio-temporal model, except for that for the northern zip, which remains low. Also note the significant increase in precision afforded by this model, which makes use of the data from all 31 days in July, 1995, instead of only that from July 15.

Figure 1.10 plots the posterior medians and upper and lower .025 quantiles produced by the spatio-temporal model by day for the ozone concentration at point A, as well as those for the block average in zip 30350. Note that the overall temporal pattern is quite similar to that for the data shown in Figure 1.9. Since point A is rather nearer to several data observation points, the confidence bands associated with it are often a bit narrower than those for the northern zip, but this pattern is not perfectly consistent over time. Also note that the relative positions of the bands for July 15 are consistent with the data pattern for this day seen in Figure 1.7, when downtown exposures were higher than those in the northern metro. Finally, the day-to-day variability in the predicted series is substantially larger than the predictive variability associated with any given day.

1.6   Summary and Discussion

In this paper we have outlined hierarchical models for handling spatially misaligned data which allow the customary Bayesian borrowing of strength across units (say, via a spatial smoothing prior), and enables full posterior inference for imputed counts at any desired level of aggregation. Our methods appear to be attractive alternatives to more ad hoc interpolation and estimation methods currently implemented in most GISs; in particular, they offer the potential for specifically estimating the combination of sampling variance and uncertainty inherent in population estimates, and using these components to make risk assessment inferences in a comprehensive manner. There are several topics for future investigation. Regarding our Section 1.2 areal data model, its application to continuous data, following the modeling suggestions preceding equation (1.1), should be explored in detail. One might also attempt an extension of our models to the spatio-temporal case. extending the nested spatio-temporal areal misalignment work by Zhu *et al.* (2000). Another important extension is the combination of misaligned areal data of the sort we have described here with point process data, such as that analyzed in Wakefield and Morris (2001). Finally, one might investigate the feasibility of our methods under higher-dimensional (e.g. three) misaligned grids, or in problems involving a multivariate response.

Turning to our Section 1.4 model for general (e.g. point-areal) misalignment, we note that our data analyses were primarily intended as illustrative, using simple exponential forms for the covariance function $c(s_i - s_{i'}; \theta)$ in Subsection 1.5.2 and $c^{(1)}(s_i - s_{i'}; \phi)$ in Subsection 1.5.3. However, other isotropic and geometrically anisotropic forms could be fitted just as easily using our approach, since they lead to only small increases in the dimension of the parameter space and the associated computational burden. Taking this idea a step further, one might wish to shed the stationarity assumption entirely. Here a rich class of nonstationary models can be created through forms $Y(s) = \sum_i a_i(s) Y_i(s)$ where the Y_i's are independent stationary processes with covariance functions c_i and the $a_i(s)$ are specified functions (Fuentes, 2000). In practice c_i would have associated parameters θ_i, whence

$$cov(Y(s), Y(s')) = \sum_i a_i(s) a_i(s') c_i(s - s'; \theta_i) . \qquad (1.29)$$

Hence in Subsection 1.4.2 (and therefore, implicitly, in Subsection 1.4.3), we can replace $c(s - s'; \theta)$ with (1.29).

Lastly, the motivation behind our point-to-zip realignment of the ozone data in Section 1.5 was our desire to relate these realigned ozone measurements to pediatric asthma emergency room (ER) visit counts which are also available at the zip code level. More specifically, and ignoring time for the moment, let Y_ℓ be the pediatric asthma ER visit count from zip ℓ, and $X(s_i)$ the average ozone level at monitoring station i. Defining the expected counts

$E_\ell = n_\ell \left(\sum_l Y_l / \sum_l n_l \right)$, we could assume

$$Y_\ell \sim Po\left(E_\ell \exp(\beta X_\ell + \Sigma_{k=1}^K \alpha_k Z_{k\ell}) \right) ,$$

where the $Z_{k\ell}$ are zip-level sociodemographic covariates (most likely socioeconomic status and race summaries). Then using our Subsection 1.4.2 approach to realign the ozone measurements, the (MC-approximated) full Bayesian model is

$$[\Pi_\ell f(Y_\ell \mid \beta, \alpha, X_\ell)] \, \hat{f}(\{X_\ell\} \mid \boldsymbol{X}_s, \mu, \sigma^2, \phi) f(\boldsymbol{X}_s \mid \mu, \sigma^2, \phi) \, p(\beta, \alpha, \mu, \sigma^2, \phi) ,$$

where $\boldsymbol{X}_s^T = (X(\boldsymbol{s}_1), \ldots, X(\boldsymbol{s}_I))$. Thus the MCMC sampler is similar to that employed above, except that it now operates over the larger space $(\alpha, \beta, \mu, \sigma^2, \phi, \{X_\ell\})$.

The extension of this model to the full spatio-temporal Poisson regression is apparent, but $X(\boldsymbol{s}_i, t_j)$ values that are missing due to failures in the ozone monitoring equipment (sometimes for days at at time) makes the computation awkward. To see this, note that we could simply include any missing $X(\boldsymbol{s}_i, t_j)$ values in the MCMC updating order, but at the price of forfeiting the Kronecker product structure used to such advantage in Subsection 1.4.3; the $X(\boldsymbol{s}_i, t_j)$ update would now require a $(IJ - 1) \times (IJ - 1)$ matrix inversion, where I is the number of ozone monitors (10) and J is the number of summer days (92). To preserve this structure, we might instead attempt to use multiple imputation approaches (Little & Rubin, 1987; Schafer, 1997) to impute values for the missing ozone observations, and then run the regression version of the spatio-temporal algorithm as before. Such imputation might itself have an iterative form, i.e.,

Impute $X(\boldsymbol{s}_i, t_j) \mid X(\boldsymbol{s}, t_j)$ (say, via kriging) ;
then impute $X(\boldsymbol{s}_i, t_j) \mid X(\boldsymbol{s}_i, t)$ (say, via an AR time series approach) .

The parameter of interest of course remains β, the effect of ozone (perhaps lagged one day) on pediatric ER asthma admissions. We hope to report the results of these and other developments in future manuscripts.

References

Beale, E.M.L. (1978). Comment on, "Curve fitting and optimal design for prediction," by A. O'Hagan. *J. Roy. Statist. Soc., Ser. B*, **40**, 25.

Bernardinelli, L., Clayton, D.G., & Montomoli, C. (1995). Bayesian estimates of disease maps: how important are priors? *Statistics in Medicine*, **14**, 2411–2431.

Bernardinelli, L. & Montomoli, C. (1992). Empirical Bayes versus fully Bayesian analysis of geographical variation in disease risk. *Statistics in Medicine*, **11**, 983–1007.

Besag, J. (1974). Spatial interaction and the statistical analysis of lattice systems (with discussion). *J. Roy. Statist. Soc., Ser. B*, **36**, 192–236.

Besag, J., Green, P., Higdon, D. & Mengersen, K. (1995). Bayesian computation and stochastic systems (with discussion). *Statistical Science*, **10**, 3–66.

Best, N.G., Ickstadt, K., Wolpert, R.L., Cockings, S., Elliott, P., Bennett, J., Bottle, A., & Reed, S. (2000). Modeling the impact of traffic-related air pollution on childhood respiratory illness (with discussion). to appear in *Case Studies in Bayesian Statistics, Volume V*, Gatsonis, C., Kass, R.E., Carlin, B.P., Carriquiry, A.L., Gelman, A., Verdinelli, I., & West, M., eds., New York: Springer-Verlag.

Brown, P., Le, N. & Zidek, J. (1994). Multivariate spatial interpolation and exposure to air pollutants. *The Canadian Journal of Statistics*, **22**, 489–509.

Carlin, B.P. & Louis, T.A. (2000). *Bayes and Empirical Bayes Methods for Data Analysis*, 2nd ed. Boca Raton, FL: Chapman & Hall/CRC Press.

Carlin, B.P., Xia, H., Devine, O., Tolbert, P. & Mulholland, J. (1999). Spatio-temporal hierarchical models for analyzing Atlanta pediatric asthma ER visit rates. In *Case Studies in Bayesian Statistics, Volume IV*, eds. Gatsonis, C. *et al.*, New York: Springer-Verlag, pp. 303–320.

Cressie, N.A.C. (1993). *Statistics for Spatial Data*, New York: John Wiley & Sons.

Cressie, N.A.C. (1996). Change of support and the modifiable areal unit problem. *Geographical Systems*, **3**, 159–180.

DeOliveira, V. (2000). Bayesian prediction of clipped Gaussian random fields. To appear *Computational Statistics and Data Analysis*.

DeOliveira, V., Kedem, B., & Short D.A. (1997). Bayesian prediction of transformed Gaussian random fields. *J. Amer. Statist. Assoc.*, **92**, 1422–1433.

Devine, O.J., Qualters, J.R., Morrissey, J.L, & Wall, P.A. (1998). Estimation of the impact of the former Feed Materials Production Center (FMPC) on Lung Cancer Mortality in the Surrounding Community. Technical report, Radiation Studies Branch, Division of Environmental Hazards and Health Effects, National Center for Environmental Health, Centers for Disease Control and Prevention.

Diggle, P.J., Tawn, J.A., & Moyeed, R.A. (1998). Model-based geostatistics (with discussion). *J. Roy. Statist. Soc., Ser. C (Applied Statistics)*, **47**, 299–350.

Ecker, M.D. & Gelfand, A.E. (1997). Bayesian variogram modeling for an isotropic spatial process. *J. Agr. Biol. Env. Statist.*, **2**, 347–369.

Fisher, P.F. & Langford M. (1995). Modeling the errors in areal interpolation between zonal systems using Monte Carlo simulation. *Environment and Planning A*, **27**, 211-224.

Flowerdew, R. & Green, M. (1989). Statistical methods for inference between incompatible zonal systems, in *Accuracy of Spatial Databases*, eds. M.F. Goodchild & S. Gopal, London: Taylor & Francis, pp. 239–248.

Flowerdew, R. & Green, M. (1991). Data integration: statistical methods for transferring data between zonal systems, in *Handling Geographical Information: Methodology and Potential Applications*, eds. Masser, I. & Blakemore, M. Essex: Longman, Harlow, pp. 38–54.

Flowerdew, R. & Green, M. (1992). Developments in areal interpolating methods and GIS. *Annals of Regional Science*, **26**, 67–78.

Flowerdew, R. & Green, M. (1994). Areal interpolation and types of data, in *Spatial Analysis and GIS*, eds. Fotheringham, S. & Rogerson, P. London: Taylor & Francis, pp. 121–145.

Flowerdew, R., Green, M., & Kehris, E. (1991). Using areal interpolation methods in Geographic Information Systems. *Papers in Regional Science*, **70**, 303–315.

Fuentes, M. (2000). A new high frequency kriging approach for nonstationary environmental processes. Technical report, Department of Statistics, North Carolina State University.

Gelman, A., Roberts, G.O., & Gilks, W.R. (1996). Efficient Metropolis jumping rules, in *Bayesian Statistics 5*, eds. Bernardo, J. M., Berger, J. O., Dawid, A. P., & A.F.M. Smith, A. F. M. Oxford: Oxford University Press, pp. 599–607.

Gelman, A. & Rubin, D.B. (1992). Inference from iterative simulation uUsing multiple sequences (with discussion). *Statistical Science*, **7**, 457–511.

Goodchild, M.F., Ansellin, L., & Deichmann, U. (1993). A framework for the areal interpolation of socioeconomic data. *Environment and Planning A*, **25**, 383-387.

Goodchild, M.F., & Lam, N.S.-N. (1980). Areal interpolation: a variant of the traditional spatial problem. *Geoprocessing*, **1**, 297-312.

Handcock, M.S. & Stein, M.L. (1993). A Bayesian analysis of kriging. *Technometrics*, **35**, 403–410.

Handcock, M.S. & Wallis, J. (1994). An approach to statistical spatial-temporal modeling of meteorological fields (with discussion). *J. Amer. Statist. Assoc.*, **89**, 368–390.

Karson, M.J., Gaudard, M., Linder, E. & Sinha, D. (1999). Bayesian analysis and computations for spatial prediction (with discussion). *Environmental and Ecological Statistics*, **6**, 147–182.

Kent, J.T. (1989). Continuity properties for random fields. *Annals of Probability*, **17**, 1432–1440.

Killough, G.G., Case, M.J., Meyer, K.R., Moore, R.E., Rope, S.K., Schmidt, D.W., Schleien, B., Sinclair, W.K., Voillequé, P.G., & Till, J.E. (1996). *Task 6: Radiation Doses and Risk to Residents from FMPC Operations from 1951-1988*, draft report, Radiological Assessments Corporation, Neeses, SC.

Lam, N.S.-N. (1983). Spatial interpolation methods: a review. *American Cartographer,* **10**, 129-149.

Langford, M., Maguire, D.J., & Unwin, D.J. (1991). The areal interpolation problem: estimating population using remote sensing in a GIS framework. in *Handling Geographical Information: Methodology and Potential Applications*, Eds. Masser, I. & Blakemore M. (Essex: Longman, Harlow) pp. 55-77.

Le, N. & Zidek, J. (1992). Interpolation with uncertain spatial covariances: a Bayesian alternative to kriging. *J. Mult. Anal.,* **43**, 351-374.

Little, R.J.A. & Rubin, D.B. (1987). *Statistical Analysis with Missing Data.* New York: Wiley.

Mardia, K.V. & Goodall, C. (1993). Spatio-temporal analyses of multivariate environmental monitoring data. In *Multivariate Environmental Statistics*, eds. Patil, G. P. & Rao, C. R. Amsterdam: Elsevier, pp. 347-386.

Mugglin, A.S. (1999). *Fully Model-Based Approaches for Spatially Misaligned Data*, unpublished PhD dissertation, Division of Biostatistics, University of Minnesota.

Mugglin, A.S. & Carlin, B.P. (1998). Hierarchical modeling in geographic information systems: population interpolation over incompatible zones. *Journal of Agricultural, Biological, and Environmental Statistics,* **3**, 111-130.

Mugglin, A.S. Carlin, B.P, Zhu, L. & Conlon, E. (1999). Bayesian areal interpolation, estimation, and smoothing: an inferential approach for Geographic Information Systems. *Environment and Planning A,* **31**, 1337-1352.

O'Hagan, A. (1978). Curve fitting and optimal design for prediction (with discussion). *J. Roy. Statist. Soc., Ser. B,* **40**, 1-42.

Rogers, J.F. & Killough, G.G. (1997). Historical dose reconstruction project: estimating the population at risk. *Health Physics,* **72**, 186-194.

Schafer, J.L. (1997). *Analysis of Incomplete Multivariate Data.* Boca Raton, FL: Chapman & Hall/CRC Press.

Stein, M.L. (1999). *Interpolation of Spatial Data: Some Theory for Kriging.* New York: Springer-Verlag.

Tobler, W.R. (1979). Smooth pycnophylactic interpolation for geographical regions (with discussion). *Journal of the American Statistical Association,* **74**, 519-536.

Tolbert, P., Mulholland, J., MacIntosh, D., Xu, F., Daniels, D., Devine, O., Carlin, B.P., Klein, M., Dorley, J., Butler, A., Nordenberg, D., Frumkin, H., Ryan, P.B., & White, M. (2000). Air pollution and pediatric emergency room visits for asthma in Atlanta. *Amer. J. Epidemiology,* **151:8**, 798-810.

Wakefield, J. & Morris, S. (2001). The Bayesian modeling of disease risk in relation to a point source, to appear *Journal of the American Statistical Association*.

Zhu, L., Carlin, B.P., English, P. & Scalf, R. (2000). Hierarchical modeling of spatio-temporally misaligned data: relating traffic density to pediatric asthma hospitalizations. *Environmetrics*, **11**, 43–61.

2 Space and Space-Time Modeling using Process Convolutions

Dave Higdon

Institute of Statistics and Decision Sciences, Duke University,
Durham NC 27708-0251, USA

Abstract. A continuous spatial model can be constructed by convolving a very simple, perhaps independent, process with a kernel or point spread function. This approach for constructing a spatial process offers a number of advantages over specification through a spatial covariogram. In particular, this *process convolution* specification leads to computational simplifications and easily extends beyond simple stationary models. This paper uses process convolution models to build space and space-time models that are flexible and able to accommodate large amounts of data. Data from environmental monitoring is considered.

2.1 Introduction

Modeling spatial data with Gaussian processes is the common thread of all geostatistical analyses. Some notable references in this area include Matheron (1963), Journel & Huijbregts (1978), Ripley (1981), Cressie (1991), Wackernagel (1995), and Stein (1999). A common approach is to model spatial dependence through the *covariogram* $c(\cdot)$, so that covariance between any two points depends only on the distance between them. Distance is typically Euclidean though other metrics are sometimes used.

An alternative, constructive, method for creating a Gaussian process over R^d is to take i.i.d. Gaussian random variables on a lattice in R^d and convolve them with an arbitrary kernel. Figure 2.1 shows a trivial example using a Gaussian kernel to convolve i.i.d. Gaussian noise. Successively increasing the density of the lattice by a factor of 2 in each dimension while reducing the variance of the variates by a factor of 2^d leads to a continuous Gaussian white

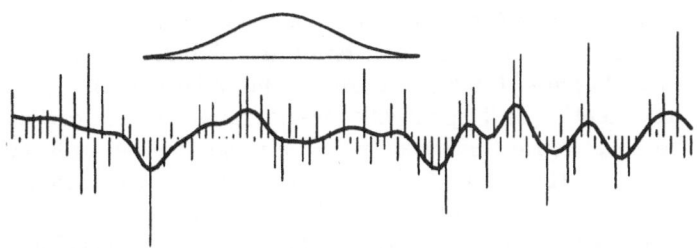

Fig. 2.1. A one-dimensional Gaussian process obtained from smoothed white noise.

noise process over R^d. The convolution of this process can be equivalently defined using some covariogram in R^d. Though defining a process by the convolution of i.i.d. Gaussian lattice variables gives very similar results to defining a process by the covariogram, the convolution construction can be readily extended to allow for non-standard features such as non-stationarity, edge effects, dimension reduction, non-Gaussian fields, and alternative space-time models. This is the primary motivation for taking this approach.

2.2   Constructing Spatial Models via Moving Averages

One may construct a Gaussian process $z(s)$ over a general spatial (and temporal) region $\mathbf{S}$, such as the real plane, by convolving a continuous white noise process $x(s)$, $s \in \mathbf{S}$ with a smoothing kernel $k(s)$ so that

$$z(s) = \int_{\mathbf{S}} k(u - s)x(u)du, \quad \text{for } s \in \mathbf{S}. \tag{2.1}$$

The resulting covariance function for $z(s)$ depends only on the displacement vector $d = s - s'$ and is given by

$$c(d) = \text{Cov}(z(s), z(s')) = \int_{\mathbf{S}} k(u-s)k(u-s')du = \int_{\mathbf{S}} k(u-d)k(u)du. \tag{2.2}$$

As a special case, if $\mathbf{S}$ is R^m and $k(s)$ is isotropic, then $z(s)$ is also isotropic, with covariance function $c(d)$ that depends only on the magnitude of d. In this case there is a one to one relationship between the smoothing kernel $k(d)$ and the covariogram $c(d)$, provided either $\int_{R^p} k(s)ds < \infty$ and $\int_{R^p} k^2(s)ds < \infty$ or $c(s)$ is integrable and positive definite. The relationship is based on the convolution theorem for Fourier transforms and is shown below,

$$k(s) \overset{FT}{\to} K(\omega) \overset{\cdot^2}{\to} K^2(\omega) \overset{IFT}{\to} c(s)$$
$$k(s) \overset{IFT}{\leftarrow} C^{\frac{1}{2}}(\omega) \overset{\sqrt{\cdot}}{\leftarrow} C(\omega) \overset{FT}{\leftarrow} c(s)$$

where FT and IFT denote the Fourier transform and its inverse, and the functions $\cdot^2$ and $\sqrt{\cdot}$ are applied pointwise. This gives the relationship between the spectrum $C(\omega)$ of a covariogram $c(s)$ and its resulting kernel $k(s)$: $C(\omega)$ is the square of the Fourier transform of $k(s)$. Note this relationship is no longer one to one if the process is not isotropic. In this case, multiple kernels can give rise to the same covariance function. The duality between the moving average process and a stationary Gaussian process determined by its variogram is explored in more detail in Thiebaux & Pedder (1985, ch. 5) and Barry & VerHoef (1996). Figure 2.2 shows kernels that give standard Gaussian, exponential and spherical covariograms for the process $z(s)$. In addition, the covariogram induced by the biweight kernel (Cleveland 1979) is also shown. Like the Gaussian kernel, it results in a covariogram that is fairly flat near the origin, and like the spherical covariogram, the dependence dies off completely after a fixed distance. This particular kernel is used in the ozone modeling examples later in this paper.

kernel	covariance function

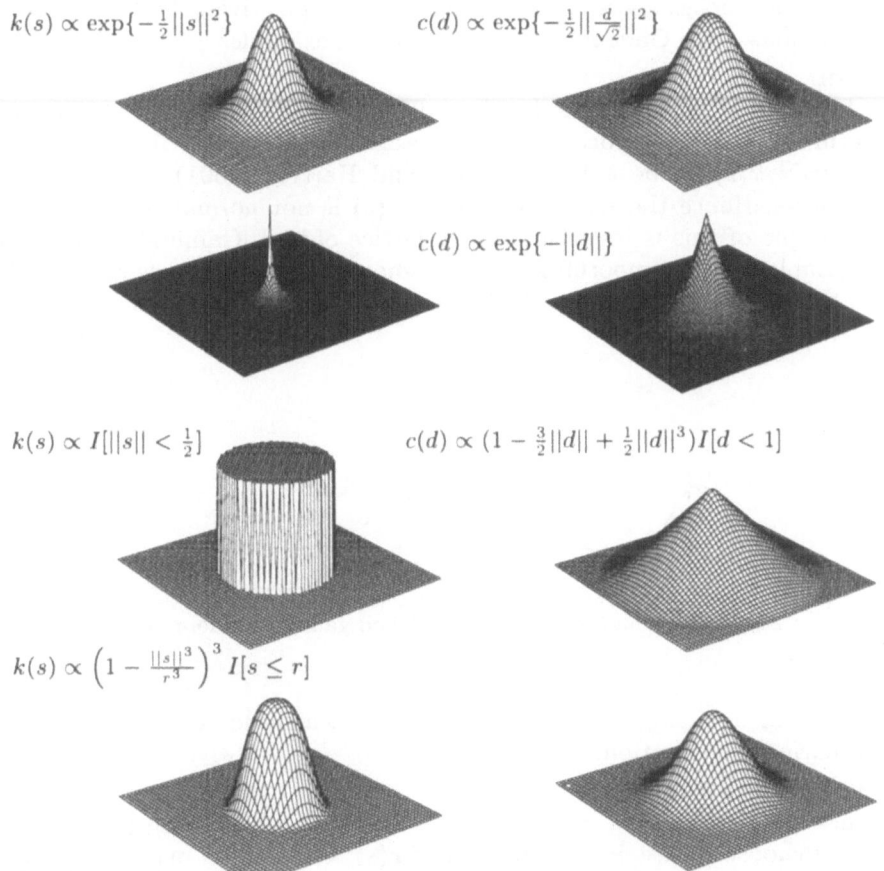

$k(s) \propto \exp\{-\frac{1}{2}\|s\|^2\}$
$c(d) \propto \exp\{-\frac{1}{2}\|\frac{d}{\sqrt{2}}\|^2\}$

$c(d) \propto \exp\{-\|d\|\}$

$k(s) \propto I[\|s\| < \frac{1}{2}]$
$c(d) \propto (1 - \frac{3}{2}\|d\| + \frac{1}{2}\|d\|^3)I[d < 1]$

$k(s) \propto \left(1 - \frac{\|s\|^3}{r^3}\right)^3 I[s \leq r]$

Fig. 2.2. Various kernels and their induced covariance functions in the two-dimensional plane.

2.2.1 Process Convolution Models

Under this moving average based approach, the model for the spatial process $z(s)$ is determined by the model specification for the latent process $x(s)$ and the smoothing kernel $k(s)$. One's choice of the latent process and smoothing kernel can lead to a number of interesting modeling approaches, a few of which are given below.

- **non-parametric covariance modeling**

  One appeal of the moving average representation is that one can model the smoothing kernel $k(s)$ rather than the covariogram $c(s)$ which must be positive definite. For example, Barry & Ver Hoef (1996), Kern (2000), and Ver Hoef *et al.* (2000) specify flexible models for $k(s)$ to build non-standard covariances for Gaussian processes. Note that there are alternative approaches specifying a flexible class of positive definite covariance functions – see Gelfand & Ecker (1997) for example.

- **non-normal** $x(s)$
  A simple extension of the basic model is to modify the specification governing the latent process $x(s)$. For example Wolpert & Ickstadt (1999) specify $x(s)$ to be a Lévy process and Hartvig (2001) uses a Strauss process. Hence the resulting process $z(s)$ is not normal. A very simple example of this is to smooth out a lattice of i.i.d. Gamma$(2,1)$ random variables with a smoothing kernel as shown below. This essentially yields a stationary non-Gaussian field. Such a modeling approach may be appro-

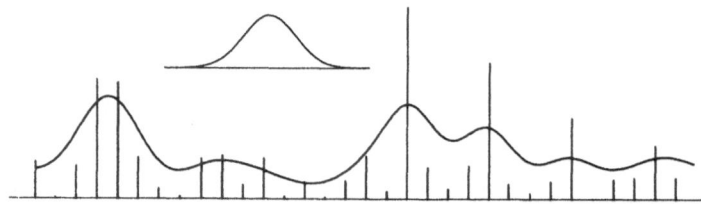

Fig. 2.3. One-dimensional smoothed gamma random field.

priate in modeling rates or concentrations which do not typically follow a normal distribution.

- **restricting the domain of** $x(s)$
  Instead of specifying $x(s)$ to be a continuous white noise process, one may choose to restrict the domain of $x(s)$. One reason may be to better account for dependencies in the system that is being modeled. For example, Kern (2000) uses such a modified process to account for irregularly shaped edges while modeling nitrogen concentrations in the Chesapeake Bay. This modification of the latent process at the edges could also be accompanied by a modification of the kernel in regions near the edges.

- **dimension reduction**
  One may choose to restrict the latent model $x(s)$ to locations $s_1, \ldots, s_m$ – over a coarse lattice for example. In this case, a small number of parameters $x(s_1), \ldots, x(s_m)$ effectively control the entire spatial process $z(s)$, even though $z(s)$ is continuous and may be required at thousands of locations. This is the main idea that I focus on for this paper. In addition to environmental monitoring problems, this approach has proven fruitful in inverse problems as well, where an economical parameterization of a spatial field can greatly facilitate computation.

- **nonstationary spatial covariance**
  The basic representation can be extended to allow the smoothing kernel

to vary with spatial location. In this case the spatial process is given by

$$z(s) = \int_S k_s(u)x(u)du$$

where $k_s(u)$ denotes a kernel which is centered at site s and whose shape also depends on s. By allowing the family of smoothing kernels $k_s(u)$, $s \in$ S to change slowly with spatial location, one can model a spatial process whose dependence structure can vary with s. Examples of using such non-stationary processes in environmental applications can be found in Higdon (1998), Higdon *et al.* (1999) and Hartvig (2001).

- **space-time models**
  By augmenting the general space S with time T so that the latent process and kernels are now defined over both space and time, $S \times T$, the basic formulation may be applied to space-time models. The latent process is defined over space and time $x(s,t)$ as is the smoothing kernel $k(s,t)$. See Higdon (1998) for an example. An intriguing alternative which is explored in this paper is to allow the latent process $x(s,t)$ to evolve over time. Along with a purely spatial kernel $k(s)$, a space time process is constructed by spatially smoothing the latent process.

$$z(s,t) = \int_S k(u-s)x(u,t)du \qquad (2.3)$$

The final example of this paper combines this idea with dimension reduction so that the specified space time model can accommodate large amounts of data.

- **building dependent spatial processes**
  Finally, the process convolution approach gives an approach to build dependent spatial processes (see Ver Hoef & Barry 1998, and Ver Hoef *et al.* 2000 for example). The basic idea is to build processes $z_j(s)$ that share part of a common latent process in their construction. For example, two processes $z_1(s)$ and $z_2(s)$ could be constructed as:

$$z_1(s) = \int_{S_0 \cup S_1} k_1(u-s)x(u)du$$

$$z_2(s) = \int_{S_0 \cup S_2} k_2(u-s)x(u)du$$

where the underlying latent process $x(s)$ resides on the union of the disjoint spaces $S = S_0 \cup S_1 \cup S_2$ and is independent on these separate subspaces. The dependence of $z_1(s)$ and $z_2(s)$ arises from their shared dependence upon $x(s)$ for $s \in S_0$. A schematic of this approach is given in Figure 2.4 below.

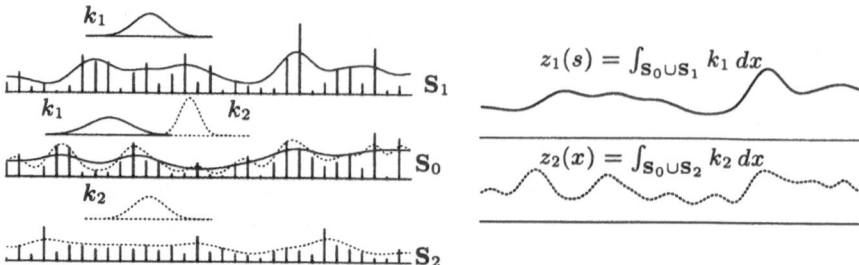

Fig. 2.4. A correlation is induced between processes $z_1(s)$ and $z_2(s)$ through common dependence upon $x(s)$ within S_0. This spatial example is readily extendible to multiple fields over space and time.

2.3   Basic Spatial Model

The goal here is to present methodology for constructing spatial models that are flexible and sufficiently tractable so that inference can be carried out with fairly large datasets. Because of this, a sparse support set for the latent process $x(s)$ is required. Hence this paper will not consider models which specify $x(s)$ to be a continuous white noise process. For such an example see Higdon *et al.*(1999).

Here we develop the formulation of the basic model. Some modifications are considered afterwards. Let $y_1, \ldots, y_n$ be data recorded over the spatial locations $s_1, \ldots, s_n$ in **S**. Perhaps the simplest spatial model represents the data as the sum of an overall mean μ, a spatial process $z = (z_1, \ldots, z_n)^T$, and Gaussian white noise $\epsilon = (\epsilon_1, \ldots, \epsilon_n)^T$ with variance σ_ϵ^2,

$$y = \mu + z + \epsilon$$

where the elements of z are the restriction of the spatial process $z(s)$ to the data locations $s_1, \ldots, s_n$.

We define $z(s)$ to be a mean zero Gaussian process. But rather than specify $z(s)$ through its covariance function, it is determined by the latent process $x(s)$ and the smoothing kernel $k(s)$. We restrict the latent process $x(s)$ to be nonzero at the spatial sites $\omega_1, \ldots, \omega_m$, also in **S** and define $x = (x_1, \ldots, x_m)^T$ where $x_j = x(\omega_j)$, $j = 1, \ldots, m$. Each x_j is then modeled as independent draws from a $N(0, \sigma_x^2)$ distribution. The resulting continuous Gaussian process is then

$$z(s) = \sum_{j=1}^{m} x_j k(s - \omega_j) \tag{2.4}$$

where $k(\cdot - \omega_j)$ is a kernel centered at ω_j. For the applications considered in this paper the smoothing kernel $k(\cdot)$ will be a radially symmetric kernel, such as a bivariate Gaussian density or any of the other kernels in Figure 2.2.

This gives the linear model

$$y = \mu 1_n + Kx + \epsilon \qquad (2.5)$$

where 1_n is the n-vector of 1's, the elements of K are given by

$$K_{ij} = k(s_i - \omega_j)x_j,$$
$$x \sim N(0, \sigma_x^2 I_m), \text{ and}$$
$$\epsilon \sim N(0, \sigma_\epsilon^2 I_n).$$

This is a basic mixed effects model. Inference can be carried out using a statistical package that uses likelihood based approaches for general mixed models such as SAS's **proc mixed** (Wolfinger *et al.* 1996) or lme in S-Plus (Pinheiro & Bates 2000). See the appendix for S-Plus 5.0 code for fitting the 1-d models here.

An alternative to straight likelihood based methods is a Bayesian approach. A fully Bayesian approach requires a prior specification for the remaining parameters μ, σ_x^2, and σ_ϵ^2. A 'default' formulation would give an improper uniform prior to μ ($\pi(\mu) \propto 1$) and rather flat gamma priors to σ_x^{-2} and σ_ϵ^{-2}. No closed form solution is available under this formulation, hence MCMC or some other approach for exploring the resulting posterior distribution will be required. If one fixes the ratio σ_x/σ_ϵ, the posterior distribution for x can be obtained in closed form (a multivariate t distribution). For applications that are data rich, it may be quite reasonable to estimate σ_x/σ_ϵ up front and then treat it as fixed.

2.3.1   Simple 1-d Example

Consider the data pairs (y_i, s_i), $i = 1, \ldots, n = 30$ shown in Figure 2.5. A process convolution model (2.5) is constructed by defining $k(s)$ to be a univariate Gaussian density with a standard deviation of 2, and defining the latent process support so that the ω_js are $m = 7$ equally spaced points ranging from -1 to 12 (as shown in the lower portion of Figure 2.5).

Note that this combination of kernel width and spacings of the ω_js yields a spatial process $z(s)$ via (2.4) that is nearly stationary. If the spacings become much larger, or if the kernel width is reduced, the covariance structure for $z(s)$ becomes unduly influenced by sparseness artifacts.

The resulting mixed effects model is fit using REML (Patterson & Thompson 1971) and the fitted values (best linear unbiased predictors) are given by the solid line in the top part of the figure. The bottom part of the figure shows the locations of the ω_js (black dots) and also shows the estimated values for each x_j. From (2.5) the fitted values can be represented as

$$\hat{y} = \hat{\mu} 1_n + \sum_{j=1}^{m} K^j \hat{x}_j$$

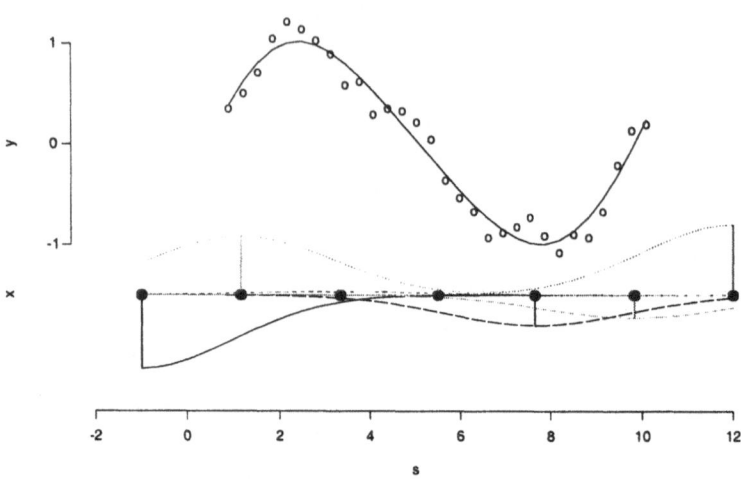

Fig. 2.5. A simulated spatial dataset with the resulting fitted values using a mixed model fit with REML (top figure). The grid locations $\omega_1, \ldots, \omega_m$ are marked by the solid dots below. The estimated values for the x_js are shown by the vertical segments at each ω_j (reduced by a factor of 3). The kernels centered at each ω_j show the 'bases' $k(s - \omega_j)x_j$ that sum to the fitted surface.

where K^j is the jth column of the matrix K of equation (2.5). The lines in the bottom part of Figure 2.5 show the terms $K^j x_j$, $j = 1, \ldots, 7$.

2.3.2   A 2-d Example

As a more serious example we consider the maximum 8 hour averaged ozone concentration over the eastern United States on a late spring day in 1999 (Figure 2.6). After looking at some empirical variograms of the data, I settled on specifying $k(s)$ to be an isotropic two dimensional tricube kernel with a range of 9 degrees. Hence the induced covariogram dies off at about 15 degrees. More detailed approaches for choosing the smoothing kernel can be found in Barry & Ver Hoef (1996), Ver Hoef & Barry (1998), and Ver Hoef *et al.* (2000). The selected ω_js are shown by the black dots in the right hand frame of Figure 2.6. They are arranged on an isotropic hexagonal grid so that each interior ω_j is equidistant to three other $\omega_{j'}$s. Again the fitted surface was obtained via REML. Note that when the support of $x(s)$ is rather sparsely spaced as it is here, the straight least squares fit is nearly identical to the REML fit. The second order properties of these fitted values have not yet been compared.

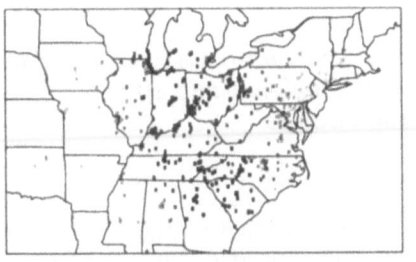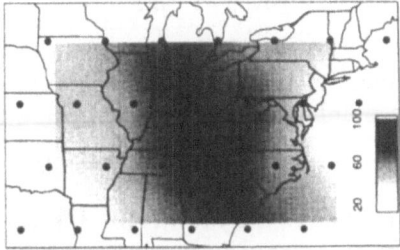

Fig. 2.6. Measured ozone concentrations and fitted concentration surface. Left: Daily maximum of the eight hour running average ozone concentration for a single day. Right: Fitted concentration surface; the dots represent the support points $\omega_1, \ldots, \omega_{27}$ of the latent process $x(s)$.

2.4   A Multiresolution Model

One can formulate a multiresolution model by representing the spatial process $z(s)$ as a sum of component processes

$$z(s) = \sum_{\ell=1}^{p} z_\ell(s)$$

where each $z_\ell(s)$ is represented through its own process convolution model determined by the pair $\{x_\ell(s), k_\ell(s)\}$, $\ell = 1, \ldots, p$. Each latent process $x_\ell(s)$ has support over spatial sites $\omega_{\ell 1}, \ldots, \omega_{\ell m_\ell}$ and we use $x_{\ell j}$ as shorthand for $x(\omega_{\ell j})$. Hence the separate processes are given by

$$z_\ell(s) = \sum_{j=1}^{m_\ell} k_\ell(s - \omega_{\ell j}) x_{\ell j}.$$

The multiresolution is captured in the kernels which become narrower as ℓ increases. Hence each additional $z_\ell(s)$ accounts for additional small scale detail.

As with the basic formulation, the model can be represented as a mixed model

$$y = \mu \mathbf{1} + \sum_{\ell=1}^{p} K^\ell x_\ell + \epsilon$$

where x_ℓ is the m_ℓ vector $(x_{\ell 1}, \ldots, x_{\ell m_\ell})^T$, the elements of K^ℓ are given by

$$K_{ij}^\ell = k_\ell(s_i - \omega_{\ell j}) x_{\ell j},$$
$$x_\ell \sim N(0, \sigma_{x_\ell}^2 I_{m_\ell}), \text{ and}$$
$$\epsilon \sim N(0, \sigma_\epsilon^2 I_n).$$

2.4.1   Simple 1-d Example

As an example, Figure 2.7 shows the $n = 30$ data points from the previous 1-d example. The spatial locations s_i vary between 1 and 10 and the actual data points were created according to the formula

$$y(s_i) = \sin(2\pi[s_i/10]) + 0.2\sin(2\pi[s_i/2.5]) + e_i$$

where the e_is are i.i.d. $N(0, .1^2)$. So the first term gives large scale variation while the second term gives a fifth of the variation at 4 times the frequency. The multiresolution model is constructed as follows:

scale (ℓ)	$k_\ell(s)$	support of $x_\ell(s)$
coarse (1)	normal; sd=2	7 sites equally spaced between -1 and 12
medium (2)	normal; sd=1	14 sites equally spaced between -1 and 12
fine (3)	normal; sd=$\frac{1}{2}$	28 sites equally spaced between -1 and 12

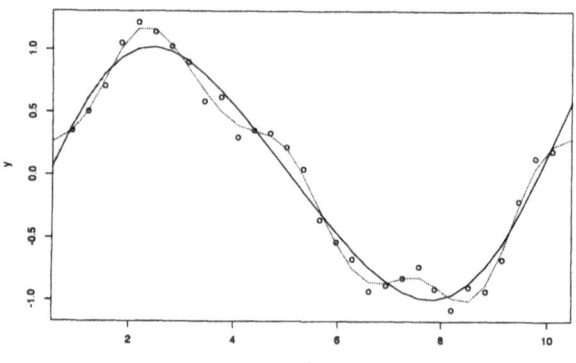

Fig. 2.7. Simulated spatial dataset with the resulting fitted values under the basic model (solid line) and the multiresolution model (dotted line). Here the multiresolution model finds the small scale structure in the data.

The fitted process is shown with the fit from the basic coarse model. Note that the multiscale formulation picks up the signal from the high frequency sine term. The estimated variance components under the two models are given below. Note that the multiresolution model picks up the high frequency sin() term at the finest resolution ($\ell = 3$) for which $k_3(s)$ nearly matches a half cycle of the sine wave.

model output for 1-d dataset

model	$\hat{\mu}$	$\hat{\sigma}_y$	$\hat{\sigma}_{x_1}$	$\hat{\sigma}_{x_2}$	$\hat{\sigma}_{x_3}$	reml logLik
basic	-0.01	0.15	8.25	$-$	$-$	0.38
multires	0.11	0.08	4.01	0.00	0.25	9.74

2.4.2   A 2-d Example

Similarly, this multiresolution model can be applied to the ozone example of the previous section. Here the multiresolution model is specified to have $p = 2$ levels. The first, coarse component ($\ell = 1$) is exactly the specification from the original ozone model so that the pair $\{x_1(s), k_1(s)\}$ is given in Section 2.3.2. The fine component refines the coarse specification by a factor of two in both spatial coordinates. Hence the process $x_2(s)$ is restricted to a denser set of points $\omega_{2,1}, \ldots, \omega_{2,87}$ shown in the bottom left frame of Figure 2.8. Along with this, the kernel $k_2(s)$ has half the scale of $k_1(s)$ so that the kernel vanishes 4.5 degrees from its center.

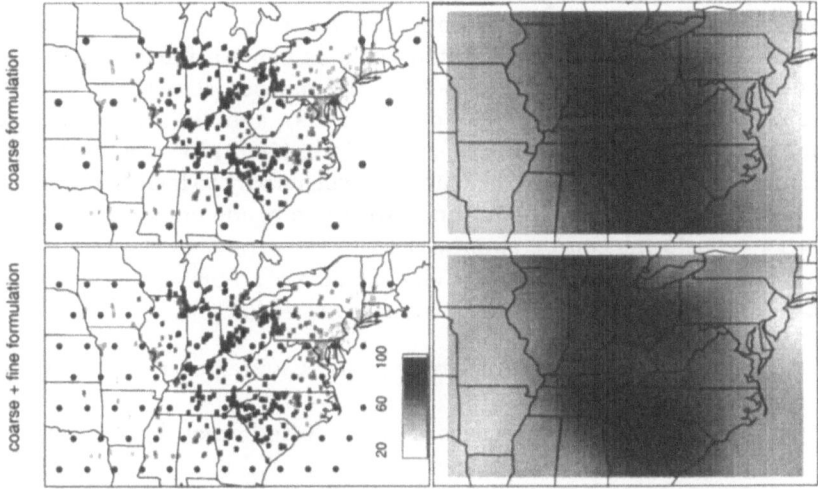

Fig. 2.8. Basic (coarse) and multiresolution models applied to the ozone data. Top left: data and location of ω_{1j}s for the latent process; Top right: fit under the basic formulation; Bottom left: data and location of the ω_{2j}s corresponding to the fine scale process – the coarse ω_{1j}s are shown in the top left figure; Bottom right: fit under the multiresolution formulation.

The data set is fit using REML and the results are summarized below. The predicted values are shown in the right hand frames of Figure 2.8. As with the 1-d example, the multiresolution fit differs from that of the basic model in the finer scale details. Such a decomposition of the ozone field is attractive since it is expected that the smaller scale fluctuations are fairly unpredictable. However the larger scale variability in the ozone concentrations does show some persistence from day to day.

model output for ozone dataset

model	$\hat{\mu}$	$\hat{\sigma}_y$	$\hat{\sigma}_{x_1}$	$\hat{\sigma}_{x_2}$	reml logLik
basic	54.33	8.04	80.35	–	−1742
multires	55.10	7.37	55.88	24.12	−1721

2.5   Building Space-Time Models

Perhaps the biggest attraction to these process convolution models is that they give a framework for developing new classes of space and space-time models that allow for more realistic space-time dependence while maintaining some analytic tractability. Generally, one can construct a space-time process by first defining a simple, possibly discrete, process over space and time, and then smoothing it out with one or more kernels, giving a smooth process over space and time.

This constructive approach is appealing since the resulting models can be extended to allow for generalizations such as non-stationarity, non-Gaussian models, and non-separable space-time dependence structures. See Wolpert & Ickstadt (1998), Ickstadt & Wolpert (1999), and Higdon *et al.* (1999) for some purely spatial applications, and Higdon (1998) for a space-time model. In addition, models can be constructed in such a way to facilitate computation – such as restricting the underlying process to reside on a lattice so that fast Fourier transforms can be employed.

For example, a simple but non-trivial example of a non-separable, non-Gaussian space-time process can be constructed as follows:

1. distribute random variables over space s and time t according to some marked point process.
2. smooth out this process according to some kernel defined over space s and time t.

This scheme for generating a realization from such a process is shown in Figure 2.9. In this particular example, i.i.d. exponential random variables are associated with each location, resulting in a strictly positive, non-Gaussian field. Note that space is only one-dimensional here to make the figure easy to understand – a 2- or 3-dimensional spatial component could also be considered.

This fairly simple example might be satisfactory for a pollutant concentration under a constant, prevailing wind. However, it may not very well account for a changing wind pattern. In such cases it might be preferable to have the underlying point process evolve over time and smooth only over the spatial component. Figure 2.10 shows a latent process defined over a spatial lattice that is slowly changing in magnitude and spatial location over time. Such a process is appealing since the spatial 'shifting' of this latent process could be linked to meteorological data for the region.

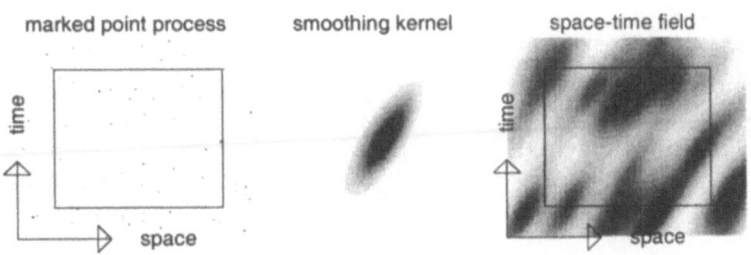

Fig. 2.9. Smoothing a marked point process defined over space and time yields a space time surface.

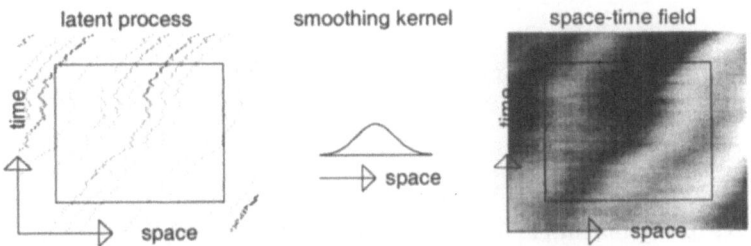

Fig. 2.10. A space time process constructed by spatially smoothing a process $x(s,t)$ whose values and support are both evolving over time.

2.5.1 A Space-Time Model for Ozone Concentrations

Using the ideas above, a space-time model for 30 consecutive days of ozone concentrations is constructed. We let $t \in \mathbf{T} = \{1, \ldots, 30\}$ index time (in days) and define the spatial support for the latent process over $\mathbf{W} = \{\omega_1, \ldots, \omega_{27}\}$ exactly as in the basic model of Section 2.3.2 so that $x(s,t)$ is nonzero over the set $\mathbf{W} \times \mathbf{T}$. Rather than define the random variables $x_{jt} = x(\omega_j, t)$ to be i.i.d., each sequence $\{x_{jt}\}$, $t = 1, \ldots, 30$ is specified to follow a Gaussian random walk. By accounting for the temporal dependence within the latent process $x(s,t)$, the induced space-time process is obtained by smoothing out $x(s,t)$ spatially

$$z(s,t) = \int_{\mathbf{S}} k(u - s)x(u,t)du$$
$$= \sum_j k(\omega_j - s)x_{jt}$$

The data consist of around 500 measurements recorded each of the 30 days. The first 8 days of measurements are shown in Figure 2.11. With so many stations, each day there are a small number that cannot report data. Hence the actual number of datapoints recorded each time step n_t varies. Conditional

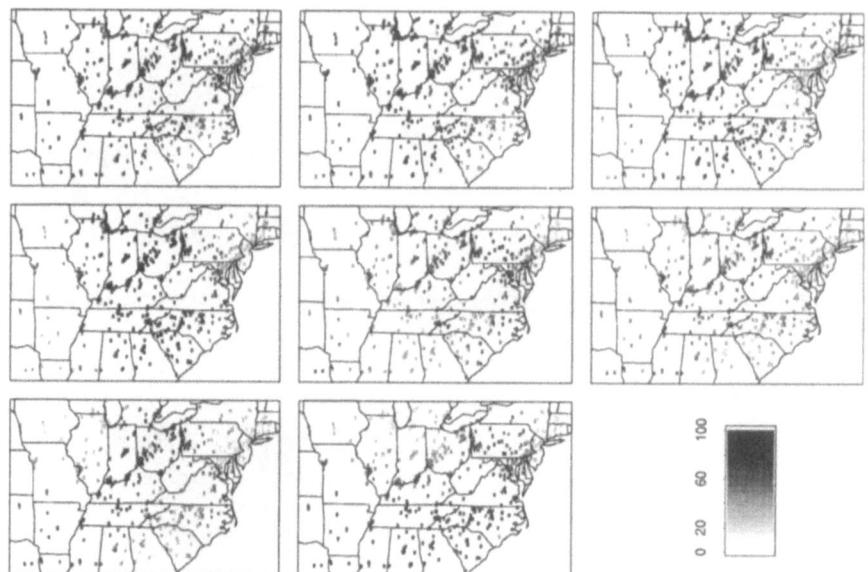

Fig. 2.11. Daily maximum for the eight hour running average of ozone concentration for eight consecutive days.

on the latent process values $x_t = (x_{1,t}, \ldots, x_{27,t})^T$, $t = 1, \ldots, 30$, a model for the data $y_t = (y_{1t}, \ldots, y_{n_t t})^T$ which is recorded at sites $s_{1t}, \ldots, s_{n_t t}$ can be expressed by the two evolution equations

$$y_t = K^t x_t + \epsilon_t \tag{2.6}$$
$$x_t = x_{t-1} + \nu_t \tag{2.7}$$

where K^t is the $n_t \times 27$ matrix given by

$$K_{ij}^t = k(s_{it} - \omega_j)x_{jt}, \ t = 1, \ldots, 30,$$
$$\epsilon_t \overset{\text{i.i.d.}}{\sim} N(0, \sigma_\epsilon^2), \ t = 1, \ldots, 30,$$
$$\nu_t \overset{\text{i.i.d.}}{\sim} N(0, \sigma_\nu^2), \ t = 1, \ldots, 30, \text{ and}$$
$$x_1 \sim N(0, \sigma_x^2 I_{27}).$$

$$\tag{2.8}$$

This model is readily amenable to the dynamic linear model (DLM) machinery of West & Harrison (1997). Other alternatives are a fully Bayesian analysis via MCMC or a REML based approach. See Stroud *et al.* (1999) for a very similar DLM based modeling approach.

For this example flat, but proper, Gamma priors were specified for the precisions $1/\sigma_\epsilon^2$, $1/\sigma_\nu^2$, and $1/\sigma_x^2$. Estimation was then carried out using MCMC.

The resulting posterior mean estimates for the fitted values are shown in Figure 2.12. In addition, Figure 2.13 shows the posterior means for $x(s,t)$ at three interior spatial support points as a function of time.

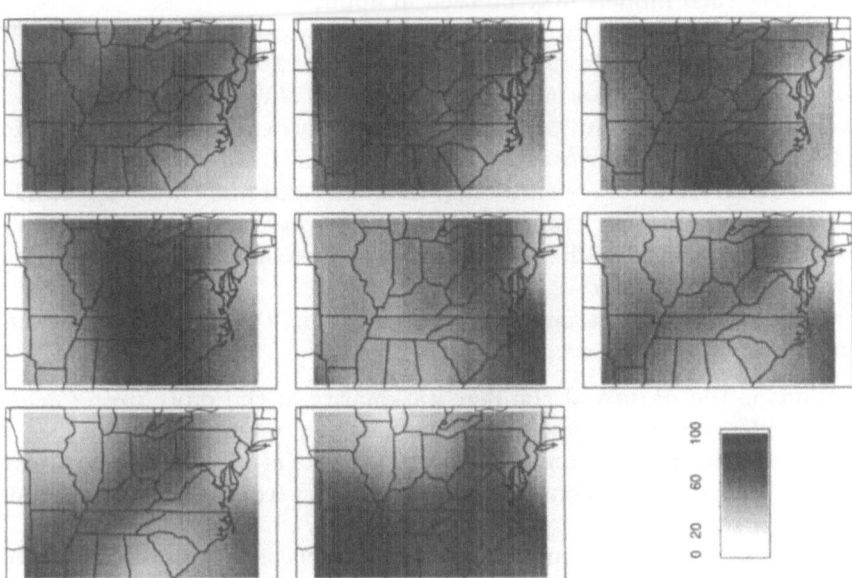

Fig. 2.12. Posterior mean surfaces of the 8 hour daily maximum ozone concentrations for eight consecutive days.

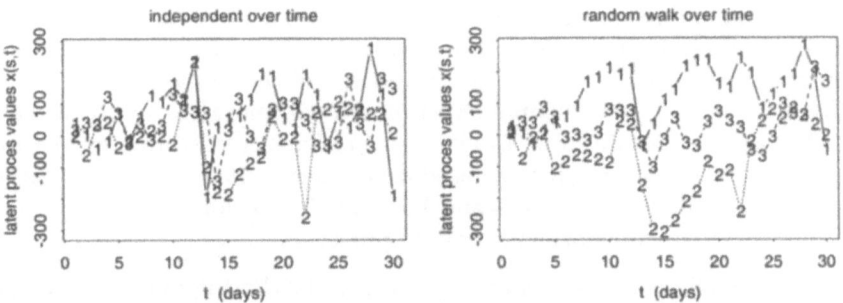

Fig. 2.13. Posterior mean estimates of the latent process $x(\omega_j, t)$, $t = 1, \ldots, 30$ at three interior locations ω_j. The left hand plot shows estimates under a model with i.i.d. x_{jt}; the right hand plot shows the estimates under the random walk formulation.

2.6   Discussion

This paper has focused on tools for building space and space-time models, rather than focusing on actual data analyses. The hope is to give readers an idea of how such models may be used in applications.

The approach is similar in spirit to using empirical orthogonal functions to reduce dimensionality in spatial fields (see von Storch & Zwiers 1999, for example), but the process convolution approach offers some advantages. Like wavelets, process convolutions allow local control over the spatial field. Interpolation is 'built-in' to process convolutions; it is less obvious how one should interpolate an EOF representation of the data. And finally, the approach can be understood through standard Gaussian process theory. Hence the vast literature regarding building Gaussian process models is also applicable to process convolution models.

Finally, I note that research in regard to modeling the ozone data is ongoing. Some issues of interest are:

- What is the nature of the non-stationarity in the daily ozone fields? If dependence does vary with spatial location, is the nature of this non-stationarity similar from day to day? Nychka *et al.* (1999) give a wavelet based approch for estimating non-stationarity from replicate observations over time.
- In the multiresolution model, how many resolution levels are appropriate? Presumably, the answer to this question depends on the actual inference problem, as well as the spatial process to be modeled. Also, the data are likely to inform only to a limited resolution. Any information about smaller scales will have to be obtained from some other source and built into the prior model. How can such fine scale information be obtained?
- It may be advantageous to combine the multiresolution model with the space-time model. It's possible that the coarse resolution component of the spatial model shows temporal dependence, while finer resolution components do not. How should one determine which resolutions should be incorporated into the temporal component of the model?
- By conditioning on a couple of estimated parameters, the posterior distribution can be obtained in closed form. Such a posterior form will be amenable to powerful utility based design methodology (Müller 1999). It will be of interest to understand how this 'conditional' posterior differs from the more appropriate marginal posterior.

We expect this line of research to play a role in furthering our understanding the nature and variability of ozone and other pollutants.

Appendix

Below is the code that produces the REML solutions for the one-dimensional examples. This was run using S-Plus 2000 for Windows.

```
# a fake dataset to make the bumps with
n_30    # of data points
m_7     # number of support sites for x(s)
# create sites s
s_seq(1,10,length=n)
# create the data y
e1_rnorm(n,sd=.1)
e2_cos(s/10*2*pi*4)*.2
y_sin(s/10*2*pi)+e2+e1
plot(s,y)

# locations of support points
w_seq(1-2,10+2,length=m)
# width of kernel
sdkern_2

# create the matrix K
K_matrix(NA,ncol=m,nrow=n)
for(i in 1:m){
  K[,i]_dnorm(s,mean=w[i],sd=sdkern)
}

# create a dataframe to hold the data
df1_data.frame(y=y,K=K,sub=1)
df1$sub_as.factor(df1$sub)

# now a fit a mixed model using lme
a1_lme(fixed= y ~ 1,
       random= pdIdent(~K-1),
       data=df1,na.action=na.omit)
# obtain and plot the fitted values
a1p_as.vector(predict(a1,df1))
lines(s,a1p,lty=1)

# now a multiscale version....
m1_c(7,14,28) # number of support points at each scale
w_list(NULL)  # list to hold the support locations
# make support locations w
for(i in 1:3) w[[i]]_seq(1-2,10+2,length=m1[i])
sdkern_c(2,1,.5) # kernel width by scale

# generate K matrices for each scale
K1_matrix(NA,ncol=m1[1],nrow=n)
for(i in 1:m1[1]) K1[,i]_dnorm(s,mean=w[[1]][i],sd=sdkern[1])
```

```
K2_matrix(NA,ncol=m1[2],nrow=n)
for(i in 1:m1[2]) K2[,i]_dnorm(s,mean=w[[2]][i],sd=sdkern[2])
K3_matrix(NA,ncol=m1[3],nrow=n)
for(i in 1:m1[3]) K3[,i]_dnorm(s,mean=w[[3]][i],sd=sdkern[3])

# create dataframe df2
df2_data.frame(y=y,K1=K1,K2=K2,K3=K3,sub=1)

# fit mixed effects model
a2_lme(fixed= y ~ 1,
       random= list(sub = pdIdent(~K1-1),sub=pdIdent(~K2-1),
              sub=pdIdent(~K3-1)),
       data=df2,na.action=na.omit)
# get predictions
a2p_as.vector(predict(a2,df2))
# plot it
plot(s,y)
lines(s,a1p,lty=1)
lines(s,a2p,lty=2)
```

References

Barry, R. P. & Ver Hoef, J. M. (1996). Blackbox kriging: spatial prediction without specifying variogram models. *Journal of Agricultural, Biological, and Environmental Statistics*, **1**, 297–322.

Cleveland, W. S. (1979). Robust locally weighted regression and smoothing scatterplots. *Journal of the American Statistical Association*, **74**, 829–836.

Cressie, N. A. C. (1991). *Statistics for Spatial Data*. Wiley-Interscience.

Gelfand, A. E. & Ecker, M. D. (1997). Bayesian variogram modeling for an isotropic spatial process. *Journal of Agricultural, Biological, and Environmental Statistics*, **2**, 347–369.

Guttorp, P., Meiring, W. & Sampson, P. D. (1994). A space-time analysis of ground-level ozone data. *Environmetrics*, **5**, 241–254.

Hartvig, N. V. (2001). A stochastic geometry model for fMRI data. *Scandinavian Journal of Statistics*, to appear.

Higdon, D. M. (1998). A process-convolution approach to modeling temperatures in the north Atlantic Ocean. *Journal of Environmental and Ecological Statistics*, **5**, 173–190.

Higdon, D. M., Swall, J. & Kern, J. C. (1999). Non-stationary spatial modeling. In *Bayesian Statistics 6. Proceedings of the Sixth Valencia International Meeting*, 761–768. Oxford University Press.

Kern, J. C. (2000). *Bayesian Process-Convolution Approaches to Specifying Spatial Dependence Structure*, unpublished PhD dissertation, Institute of Statistics & Decision Sciences, Duke University, Durham, USA.

Ickstadt, K. & Wolpert, R. L. (1999). Spatial regression for marked point processes. In *Bayesian Statistics 6. Proceedings of the Sixth Valencia International Meeting*, 323–341.

Littell, R. C., Milliken, G. A., Stroup, W. W. & Wolfinger, R. D. (1996). *SAS System for Mixed Models*. SAS Institute.

Matérn, B. (1986). *Spatial Variation (Second Edition)*. Springer-Verlag.

Meiring, W., Guttorp, P. & Sampson, P. D. (1998). Space-time estimation of grid-cell hourly ozone levels for assessment of a deterministic model. *Environmental and Ecological Statistics*, **5**, 197–222.

Müller, P. (1999). Simulation based optimal design. In *Bayesian Statistics 6. Proceedings of the Sixth Valencia International Meeting*, 323–341.

Nychka, D., Wikle, C. & Royle, J. A. (1999). Large spatial prediction problems and nonstationary random fields. Technical report, National Center for Atmospheric Research.

Patterson, H. D. & Thompson, R. (1971). Recovery of inter-block information when block sizes are unequal. *Biometrika*, **58**, 545–554.

Pinheiro, J. & Bates, D. (2000). *Mixed Effects models in S and S-plus*. New York: Springer.

Sampson, P. D. & Guttorp, P. (1992). Nonparametric estimation of nonstationary spatial covariance structure. *Journal of the American Statistical Association*, **87**, 108–119.

Stein, M. (1999). *Interpolation of Spatial Data: Some Theory for Kriging*. New York: Springer-Verlag.

Stroud, J., Müller, P. & Sanso, B. (1999). Dynamic models for spatio-temporal data. Technical Report 99-20, Institute of Statistics and Decision Sciences, Duke University.

Thiébaux, H. J. (1997). The power of the duality in spatial-temporal estimation. *Journal of Climatology*, **10**, 567–573.

Thiébaux, H. J. & Pedder, M. A. (1987). *Spatial Objective Analysis: with applications in atmospheric science*. San Diego: Academic Press.

Ver Hoef, J. & Barry, R. P. (1998). Constructing and fitting models for cokriging and multivariable spatial prediction. *Journal of Statistical Planning and Inference*, **69**, 275–294.

Ver Hoef, J., Cressie, N. & Barry, R. (2000). Flexible spatial models based on the fast Fourier transform (FFT) for cokriging. Technical report, Department of Statistics, The Ohio State University.

von Storch, H. & Zwiers, F. W. (1999). *Statistical Analysis in Climate Research*. New York: Cambridge University Press.

Wackernagel, H. (1995). *Multivariate Geostatistics. An Introduction With Applications*. Springer-Verlag.

West, M. & Harrison, J. (1997). *Bayesian Forecasting and Dynamic Models (Second Edition)*. New York: Springer-Verlag.

Wolpert, R. L. & Ickstadt, K. (1998). Poisson/gamma random field models for spatial statistics. *Biometrika*, **85**, 251–267.

Yaglom, A. M. (1987). *Correlation Theory of Stationary and Related Random Functions. Volume I: Basic Results*. Springer-Verlag, New York.

Yaglom, A. M. (1987). *Correlation Theory of Stationary and Related Random Functions. Volume II: Supplementary Notes and References*. Springer-Verlag, New York.

3 Multivariate Kriging for Interpolating with Data from Different Sources

H. Wackernagel[1], L. Bertino[1], J. P. Sierra[2], and J. González del Río[3]

[1] Centre de Géostatistique, Ecole des Mines de Paris,
    35 rue Saint Honoré, F-77305 Fontainebleau, France
[2] Laboratori d'Enginyeria Marítima, Universitat Politècnica de Catalunya,
    C/Jordi Girona 1-3, Campus Nord Mòdul D-1, E-08034 Barcelona, Spain
[3] Laboratorio de Tecnologías del Medio Ambiente, Universidad Politécnica de
    Valencia, Camino de Vera s/n, E-46020 Valencia, Spain

Abstract. A discussion on some aspects of multivariate kriging (cokriging, external drift method) is proposed in the first part, insisting on questions of cokriging neighborhood. Special emphasis is given to a recent discussion about collocated and multicollocated cokriging in the geostatistical literature.
The second part is unrelated to the first one, combining unique neighborhood data from two different sources in a case study in the Ebro river estuary (Spain): sounder-measured conductivity, and water samples analyzed for chlorophyll and salinity.

3.1 Introduction

Kriging is a special type of optimal linear prediction applied to random functions in space or time with the particular requirement that their covariance structure should be known. The covariance structure is described using either covariance functions, variograms or generalized covariances for a set of variables. When cross-dependence is included in the form of cross-covariance functions or cross-variograms, the corresponding multivariate kriging is termed *cokriging*. The drift, a mean function describing systematic aspects of a spatial phenomenon, can either be included into kriging as translation-invariant polynomial drift or as *external drift*, the latter being an alternate way of rendering kriging multivariate.

The configuration of the sampling sites of different types of measurements in a specific spatial/temporal domain is of basic importance: the question whether (or not) sites are shared by different measurement types will influence the formulation of the multivariate kriging approach. The *neighborhood* is the subset of data used in cokriging. How should the cokriging neighborhood be selected? How does the covariance structure come into play? These are some questions of interest that will be examined in the first part of this paper.

The second part is devoted to a case study set in the Ebro estuary (Spain) in which cokriging is applied to chemical/biological measurements on water samples taking advantage of more extensive data provided by a physical measurement device.

3.2   Data from Different Sources

Multivariate data often stem from different sources. A typical environmental example is remote sensing data and ground data taken in the same area. A more general example, demonstrated in the case study of the second part of the paper, is that of chemical measurements on one hand, which are expensive and few in number, and physical measurements on the other hand, which may be cheaper, easier to implement and with greater spatial coverage. There is usually much to gain by using physical variables to improve the estimation of chemical variables.

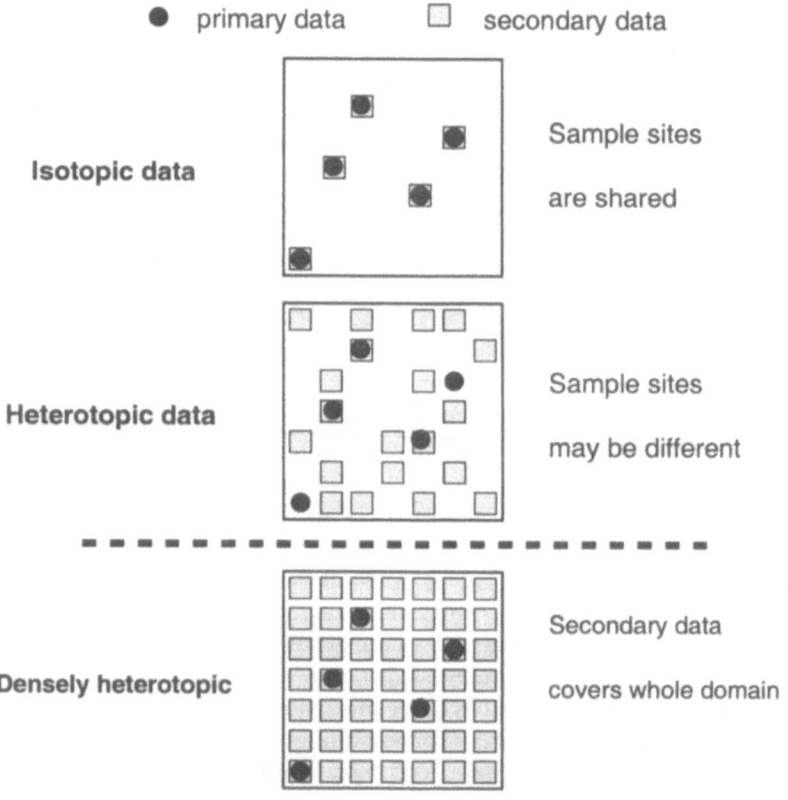

Fig. 3.1. Data location

The configuration of sample locations for different variables in the domain is of importance. Figure 3.1 displays three possible cases using a terminology proposed by Matheron (1982),

- isotopic case: the sample locations of the different variables are identical, as shown on the upper graph,

- heterotopic case: the sample locations may be different, as sketched on the middle graph.

A special case we shall call *dense secondary variable* is shown on the lower graph: when a secondary variable is available on a dense grid covering the whole domain (e.g. remote sensing or other geophysical data). A consequence of importance is that there is data about the secondary variable at every estimation location, a fact that will be exploited after presenting cokriging and the external drift method.

3.3 Simple and Ordinary Cokriging

Simple cokriging requires knowledge of the first two moments for a set of random functions. The two moments are the means m_i of the variables $Z_i(x)$, $i = 1 \ldots N$ as well as the cross-covariance functions $C_{ij}(h) = \text{cov}(Z_i(x), Z_j(x+h))$, where x and $x+h$ are two locations in the domain, and translation invariance with respect to the vector h is assumed.

Let i_0 be the index of the variable of interest. The simple cokriging estimator for predicting at an arbitrary location x_0 using data at sample locations x_α, $\alpha = 1 \ldots n_i$ (with the number of samples n_i depending on the index i) is,

$$Z_{i_0,\text{SK}}^{\star\star}(x_0) = m_{i_0} + \sum_{i=1}^{N} \sum_{\alpha=1}^{n_i} w_\alpha^i \left(Z_i(x_\alpha) - m_i \right). \tag{3.1}$$

We use a double star notation to indicate that the estimator is a cokriging, reserving a single star for a linear combination of the primary data only. The weights w_α^i are obtained by minimizing the estimation variance,

$$\text{var}\left(Z_{i_0,\text{SK}}^{\star\star}(x_0) - Z_{i_0}(x_0) \right), \tag{3.2}$$

and solving the resulting cokriging system,

$$\begin{pmatrix} C_{11} & \ldots & C_{1j} & \ldots & C_{1N} \\ \vdots & \ddots & & & \vdots \\ C_{i1} & & C_{ii} & & C_{iN} \\ \vdots & & & \ddots & \vdots \\ C_{N1} & \ldots & C_{Nj} & \ldots & C_{NN} \end{pmatrix} \begin{pmatrix} w_1 \\ \vdots \\ w_i \\ \vdots \\ w_N \end{pmatrix} = \begin{pmatrix} c_{1i_0} \\ \vdots \\ c_{ii_0} \\ \vdots \\ c_{Ni_0} \end{pmatrix}, \tag{3.3}$$

where the left hand side matrix is built up with square symmetric $n_i \times n_i$ blocks C_{ii} on the diagonal and with rectangular $n_i \times n_j$ blocks C_{ij} off the diagonal, knowing that $C_{ij} = C_{ji}^\top$. The blocks C_{ij} contain the covariances $C_{ij}(x_\alpha - x_\beta)$ between sample points for a fixed pair of variables. The vectors c_{ii_0} list the covariances with the variable of interest, for a specific variable of the set, between sample points and the prediction location. The vectors w_i contain the weights attached to the samples of each variable.

The ordinary cokriging predictor is a linear combination of weights w_α^i with data $Z_i(\boldsymbol{x}_\alpha)$ at sample locations,

$$Z_{i_0,\text{OK}}^{\star\star}(\boldsymbol{x}_0) = \sum_{i=1}^{N} \sum_{\alpha=1}^{n_i} w_\alpha^i \, Z_i(\boldsymbol{x}_\alpha). \qquad (3.4)$$

Contrarily to the simple cokriging predictor it does not require specification of means and can be applied to non-stationary phenomena characterized by cross-variograms,

$$\gamma_{ij}(\boldsymbol{h}) = \frac{1}{2} \operatorname{cov} \left(Z_i(\boldsymbol{x}+\boldsymbol{h}) - Z_i(\boldsymbol{x}), Z_j(\boldsymbol{x}+\boldsymbol{h}) - Z_j(\boldsymbol{x}) \right), \qquad (3.5)$$

assuming generally that the expectation of increments $Z_i(\boldsymbol{x}+\boldsymbol{h}) - Z_i(\boldsymbol{x})$ is zero for any translation of an arbitrary vector $\boldsymbol{h}$.

The greater generality of the ordinary cokriging predictor has to be secured by including in the cokriging system the following constraints on the weights (and corresponding Lagrange multipliers λ_i),

$$\sum_{\alpha=1}^{n_i} w_\alpha^i = \delta_{ii_0} = \begin{cases} 1 & \text{for } i = i_0, \\ 0 & \text{otherwise,} \end{cases} \qquad (3.6)$$

both for avoiding bias and for ensuring the existence of error expectation and variance.

The ability of the cross-variograms to model non-stationary phenomena (without drift) has to be balanced against the fact that they are even functions, $\gamma_{ij}(\boldsymbol{h}) = \gamma_{ji}(\boldsymbol{h})$, so that unlike cross-covariance functions, delays in time or spatial shifts between variables cannot be modeled. A different approach is advocated by Cressie (1993) and Myers (1991), the pseudo cross-variogram

$$\pi_{ij}(\boldsymbol{h}) = \frac{1}{2} \operatorname{var} \left(Z_i(\boldsymbol{x}+\boldsymbol{h}) - Z_j(\boldsymbol{x}) \right), \qquad (3.7)$$

whose practical implementation requires that the expectation of cross-increments $Z_i(\boldsymbol{x}+\boldsymbol{h}) - Z_j(\boldsymbol{x})$ is assumed to be zero, a hypothesis that has to be handled with caution when $Z_i(\boldsymbol{x})$ and $Z_j(\boldsymbol{x})$ are different types of measurements. Situations can easily be found in which the hypothesis of stationary direct increments is plausible whereas the stationarity assumption about cross-increments is unrealistic.

3.4   Cokriging with External Drifts

We may consider a second set of variables z_k, $k= 1, \ldots, K$ that are available both at the sample locations and at the prediction locations (typically the nodes of a grid spanned over the domain). We denote them in lower case as we shall not consider them as random.

Assuming the variables $Z_i(\boldsymbol{x})$ are related to the second set of variables z_k by the relations:

$$E[Z_i(\boldsymbol{x})] = \sum_{k=1}^{K} a_k^i + b_k^i z_k(\boldsymbol{x}) \qquad \text{for} \quad i = 1, \ldots, N, \tag{3.8}$$

the samples of the second set can be included into the ordinary cokriging system using additional constraints,

$$\sum_{\alpha=1}^{n_i} w_\alpha^k \, z_k(\boldsymbol{x}_\alpha) = \delta_{ii_0} \, z_k(\boldsymbol{x}_0). \tag{3.9}$$

The cokriging system with multiple external drift finally comes as

$$
\begin{cases}
\displaystyle\sum_{j=1}^{N} \sum_{\beta=1}^{n_j} w_\beta^j \, \gamma_{ij}(\boldsymbol{x}_\alpha - \boldsymbol{x}_\beta) + \lambda_i + \sum_{k=1}^{K} \lambda_k \, z_k(\boldsymbol{x}_\alpha) = \gamma_{ii_0}(\boldsymbol{x}_\alpha - \boldsymbol{x}_0) & \text{for } \forall \alpha, i, \\[2ex]
\displaystyle\sum_{\beta=1}^{n_i} w_\beta^i = \delta_{ii_0} & \text{for } \forall i \\[2ex]
\displaystyle\sum_{\beta=1}^{n_i} w_\beta^i \, z_k(\boldsymbol{x}_\beta) = \delta_{ii_0} \, z_k(\boldsymbol{x}_0) & \text{for } \forall i, k,
\end{cases}
\tag{3.10}
$$

where λ_i, λ_k are Lagrange multipliers.

The external drift variables z_k on one hand need to be linearly related to the different variables Z_i in conjunction with which they are used. On the other hand the different drift variables have to be linearly independent among themselves. A constant external drift for example is redundant with the condition that the weights of the principal variable should satisfy $\sum w_\alpha^{i_0} = 1$ and will thus cause the cokriging system to be singular.

3.5 Cokriging Neighborhood

Cokriging with many variables using all data easily generates a very large linear system to solve. This means that the choice of a subset of data around a given estimation location, called a *neighborhood*, is a crucial step in cokriging. It is of particular importance to know when, due to the particular structure of a coregionalization, the full cokriging with all data is actually equivalent to a cokriging using a subset of data, so that the neighborhood can be reduced a priori and the cokriging system simplified accordingly, thus reducing in the end the numerical effort to a considerable extent.

For isotopic data the most important aspect is to know whether the coregionalization shows direct and cross variograms that are proportional to one direct variogram which entails the cokriging to be equivalent to a separate kriging, leaving out the secondary variables. This topic will not be re-exposed here: see e.g. Wackernagel (1998) and Chilès & Delfiner (1999). Concerning heterotopic data we will focus on a case that has attracted most attention recently as it is increasingly frequently encountered in applications: the case of a dense secondary variable.

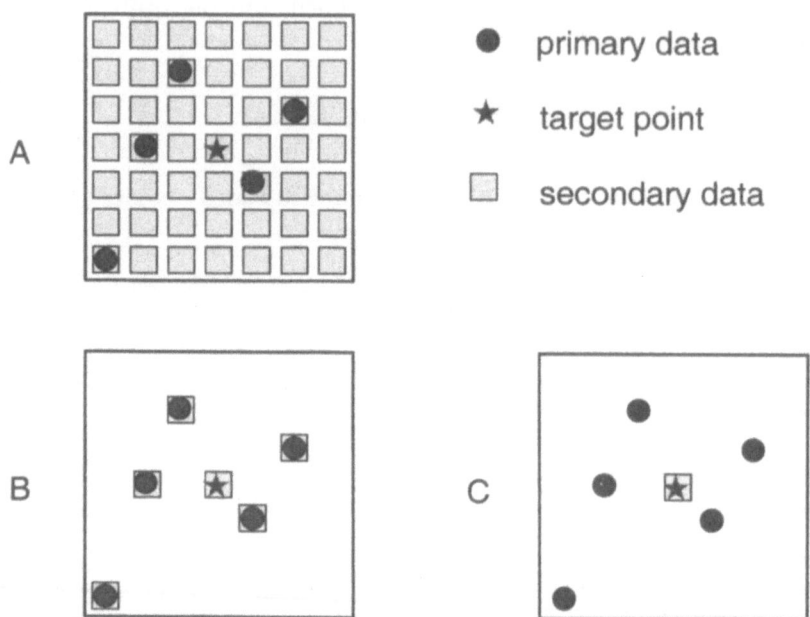

Fig. 3.2. Neighborhoods with a dense secondary variable

3.5.1   Neighborhoods with a Dense Secondary Variable

Figure 3.2 sketches three different neighborhoods for a given central estimation location (denoted by a star), primary data (denoted by full circles) as well as three alternate subsets of data from a secondary variable (denoted by squares). The neighborhood:

A uses all data available for the secondary variable,
B restricts the secondary information to the subset of locations where primary data is available as well as to the estimation location,
C merely includes a sample value of the secondary variable at the estimation location.

Case A can be termed the *full* neighborhood, while case C was called a *collocated* neighborhood by Xu *et al.* (1992) as the secondary data is collocated with the estimation location, whereas case B was termed a *multicollocated* neighborhood by Chilès & Delfiner (1999) as additionally the secondary data is also collocated with the primary data.

Using the full neighborhood with secondary data dense in space will easily lead to linear dependencies for neighboring samples in the cokriging system, causing it to be singular. The size of the system can also be numerically challenging. Vargas-Guzman & Yeh (1999) suggest a way out of numerical difficulties by starting from a small neighborhood and progressively extending the neighborhood in the framework of what they call a sequential cokriging. The collocated neighborhood is only valid for simple cokriging because using a single sample for the secondary variable in ordinary cokriging leads to a trivial cokriging weight for that sample due to the constraint. For simple cokriging the approach solely requires the inference of a correlation coefficient instead of a cross-covariance function, a simplification that can only be meaningful for coregionalization models with proportionalities. However, as shown in Wackernagel (1998, p165) the full simple cokriging with such a model does not reduce to a collocated simple cokriging.

The multicollocated neighborhood is examined in detail by Rivoirard (2000) who has shown in the bivariate case that cokriging a primary variable Z_1 with a full neighborhood is equivalent to cokriging with a multicollocated neighborhood when the covariance structure is of the type:

$$\begin{cases} C_{11}(\boldsymbol{h}) & = p^2\, C(\boldsymbol{h}) \,+\, C^R(\boldsymbol{h}), \\ C_{22}(\boldsymbol{h}) & = C(\boldsymbol{h}), \\ C_{12}(\boldsymbol{h}) & = p\, C(\boldsymbol{h}), \end{cases} \tag{3.11}$$

where p is a proportionality coefficient. Implications of this model are that the covariance function $C_{22}(\boldsymbol{h})$ is more regular than $C_{11}(\boldsymbol{h})$, if $C^R R(\boldsymbol{h})$ is less regular than $C(\boldsymbol{h})$.

The estimator of bivariate multicollocated cokriging is

$$Z_1^{\star\star}(\boldsymbol{x}_0) = Z_1^{\star}(\boldsymbol{x}_0) + p\,(Z_2(\boldsymbol{x}_0) - Z_2^{\star}(\boldsymbol{x}_0))\,, \tag{3.12}$$

where the two krigings $Z_1^{\star}$ and $Z_2^{\star}$ are both obtained using the weights from an ordinary kriging system set up with the covariance function $C^R(\boldsymbol{h})$ for the n_1 samples of the primary variable:

$$Z_1^{\star}(\boldsymbol{x}_0) = \sum_{\alpha=1}^{n_1} w_\alpha^R Z_1(\boldsymbol{x}_\alpha), \qquad Z_2^{\star}(\boldsymbol{x}_0) = \sum_{\alpha=1}^{n_1} w_\alpha^R Z_2(\boldsymbol{x}_\alpha). \tag{3.13}$$

Rivoirard (2000) obtained this result for Gaussian random functions with conditional independence of $Z_1(\boldsymbol{x}_0)$ and $Z_2(\boldsymbol{x})$ knowing $Z_2(\boldsymbol{x}_0)$. It is straightforward to derive it more generally for second-order stationary random functions simplifying the cokriging system on starting directly from the covariance function model (3.11).

This result can be generalized to the multivariate case for uncorrelated secondary variables (e.g. principal components of remote sensing channels). Let the first variable be the variable of interest, let $j = 2, \ldots, N$ be the indices of $N-1$ principal components with covariance functions $C_{jj}(h)$ and assume the coregionalization model,

$$
\begin{cases}
C_{11}(h) = \sum_{j=2}^{N} p_j{}^2 \; C_{jj}(h) + C^R(h), \\
C_{1j}(h) = p_j \; C_{jj}(h) & \text{for } j = 2, \ldots, N, \\
C_{jj'}(h) = 0 & \text{for } j \neq j',
\end{cases}
\tag{3.14}
$$

where p_j are coefficients. Knowing that remote sensing channels in our experience are not seldom intrinsically correlated (Lindner & Wackernagel 1993, Chica-Olmo & Abarca-Hernandez 1998) it makes sense to compute their principal components and this type of coregionalization model could be applied in that context.

The estimator of multivariate multicolocated cokriging is then

$$
Z_1^{\star\star}(x_0) = Z_1^{\star}(x_0) + \sum_{j=2}^{N} p_j \left(Z_j(x_0) - Z_j^{\star}(x_0) \right),
\tag{3.15}
$$

reducing the cokriging to a linear combination of coefficients p_j with differences between the known values $Z_j^{\star}(x_0)$ and the ordinary krigings $Z_j(x_0)$ computed from the n_1 sample locations of Z_1 using $C^R(h)$.

3.6   Ebro Case Study

The Ebro is the second largest river of Spain. We shall use data collected by the Polytechnic Universities of Barcelona and Valencia, on the 5th October 1999 during the third campaign of the PIONEER project. The measurements were performed between 11am and 6pm, navigating upstream from the estuary. Details on the campaign and on the interpretation of the data are given in Sierra et al. (2000a, 2000b), and González et al. (2000). We shall use these data only for demonstrative purpose, to discuss some problems in the application of kriging, cokriging and conditional simulations.

3.6.1   Kriging Conductivity

Conductivity was measured employing a multiparametric sounding Hydrolab Surveyor III with the aim of locating the freshwater-seawater interface. The measurements were performed at five locations along the river, sampling vertically with a 10 centimeter spacing. This resulted in a total of 185 conductivity values. A plot of the five profiles is shown on Figure 3.3 using symbols proportional to the value of conductivity. Conductivity expresses the salinity of the water. The transition zone between freshwater and seawater is easily

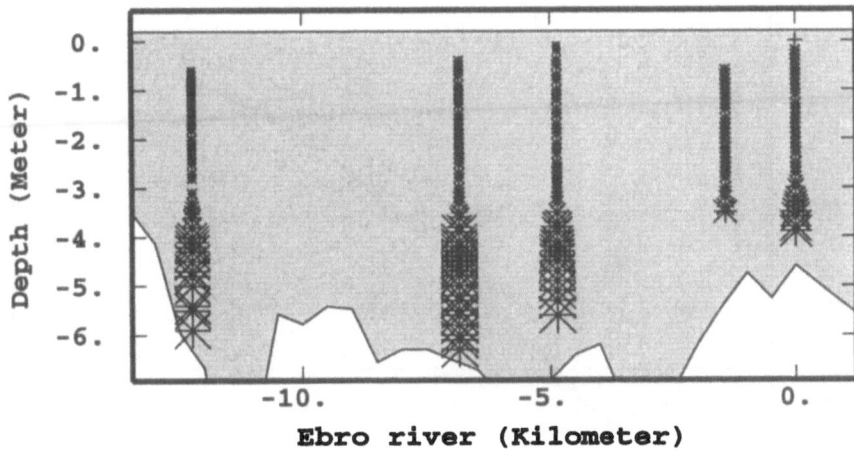

Fig. 3.3. The 185 Hydrolab Surveyor III sample locations are plotted with symbols proportional to measured conductivity

identified between 3 and 4 meter depth. The abscissa indicates the distance from the mouth of the Ebro river in kilometers.

The river bed displayed on Figure 3.3 is actually based on bathymetric measurements stemming from the second PIONEER campaign in the month of July 1999. We can assume that the bottom did not experience great changes, but obviously, if there are different river discharges, the water levels (and as a consequence the depths) will be different. We made a fast computation of these differences using the following approach: the average discharge during the July campaign was 129 m^3/s, while it amounted to 184 m^3/s in the October campaign (measured at a station upstream in the city of Tortosa). For this difference of river discharge, the water level is evaluated as about 20 cm higher on average and the bathymetry was corrected accordingly.

We will consider the problem of interpolating by kriging the conductivity profiles. Experimental variograms were computed between and within the profiles, using 60 lags of 10cm in the vertical and 100m in the horizontal. They are shown on Figure 3.4. In the horizontal direction (with a tolerance of ± 45 degrees) the variogram is denoted D1, while in the vertical it is denoted D2. Note the difference in scale in horizontal (kilometers) and vertical (meters) directions.

Considering that we know the variogram structure well only in the vertical direction we can do nothing better than to adopt the same model in the horizontal using a geometrical anisotropy. The main axes of the anisotropy ellipse are taken parallel to the horizontal and the vertical. The cubic model (M2) fitted in the vertical with a range of 7.5m was adjusted to the horizontal with a range of 17km. A nugget-effect of 2 (mS/cm)^2 was added to reflect measurement uncertainty and the sill of the cubic model is 1150 (mS/cm)^2,

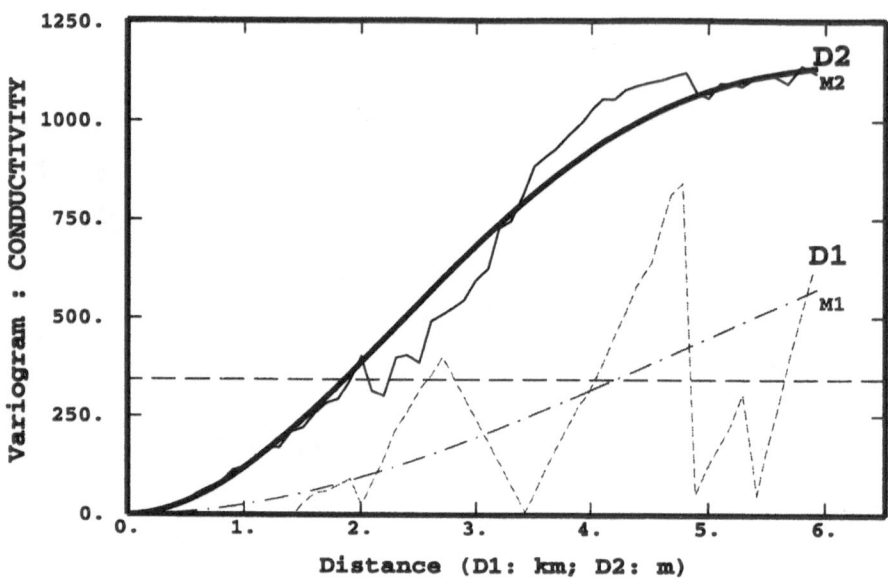

Fig. 3.4. Experimental variogram of conductivity in two directions (D1: horizontal, D2: vertical). Model variogram in both directions (M1, M2). The abscissa should be read as kilometers for the horizontal and as meters for the vertical

about three times the variance which is represented as a horizontal dotted line on Figure 3.4. The smooth parabolic behavior at the origin of the cubic model may be interpreted as reflecting the averaging over a non-point support by the physical measurement device.

An interpolation grid of 137 × 75 nodes with 100m × 10cm cells was defined, starting from an origin at (-12.9km, -6.8m). This interpolation grid and a neighborhood including all data (unique neighborhood) will be used in all examples.

The ordinary kriging of conductivity using the cubic model and filtering the nugget effect is shown on Figure 3.5. The map represents well the two phases, freshwater and seawater, suggested by the data on Figure 3.3. This picture of the spatial distribution of chlorophyll relies heavily on the geometric anisotropy built into the geostatistical model, which emphasizes the horizontal dependence between the profiles.

3.6.2   Cokriging of Chlorophyll

The Hydrolab device has been used to obtain quickly an indication about the depth of the freshwater/seawater interface. Water samples were obtained using a new device called SWIS (Salt Wedge Interface System) developed jointly by the Polytecnic Universities of Barcelona and Valencia. It consists

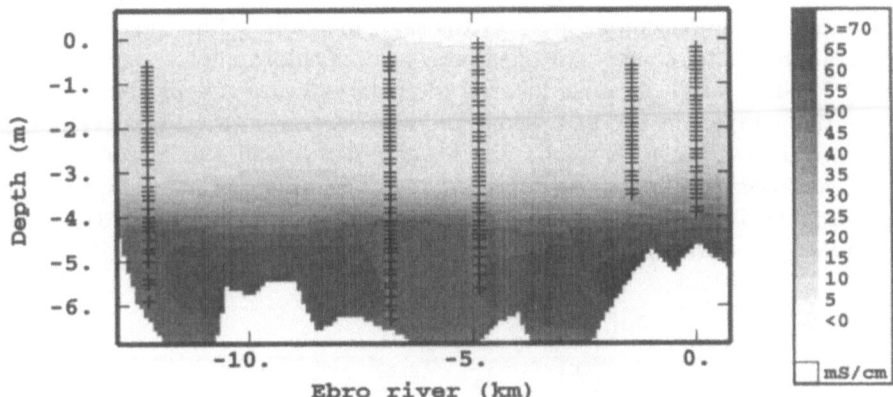

Fig. 3.5. Map of conductivity obtained by ordinary kriging, filtering the nugget effect component

of six tubes connected to a vacuum system that can be operated from the surface. Spacing the tubes at 10cm from each other, the roughly half a meter wide interface can be sampled with up to six samples in one go. Additional samples (at zero, 1.5, 3 and 4.5 meter depth) were taken, leading to a total of 47 samples for the five measurement points along the Ebro river. The number of locations where both water samples and Hydrolab samples are available is 31.

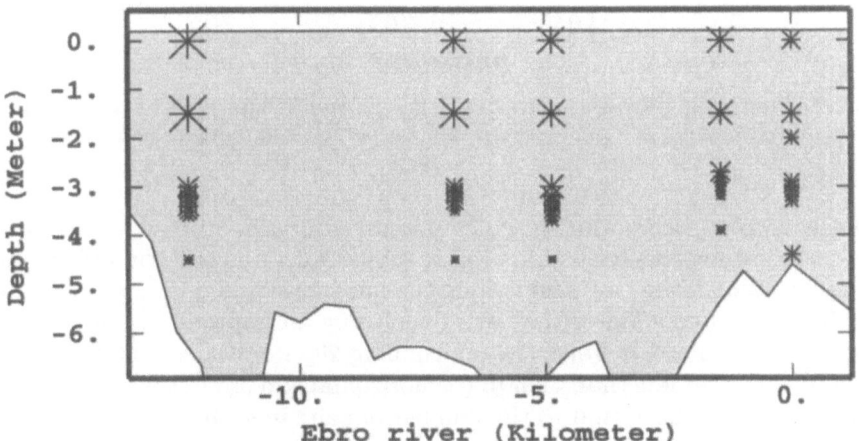

Fig. 3.6. The 47 water sample locations are plotted with symbols proportional to chlorophyll

On Figure 3.6 the water sample locations are plotted using symbols proportional to the value of chlorophyll. At the same locations salinity measurements are available. The scatter plot of salinity against chlorophyll is shown on Figure 3.7. Nine water samples located in the freshwater are plotted as stars, while the samples in the salt wedge are represented with crosses. While the relationship within both media can be assumed linear, this is not the case when considering all data. As cokriging requires a linear relationship between the variable of interest and the auxiliary variables, the logarithm (basis 10) was taken for both salinity and conductivity.

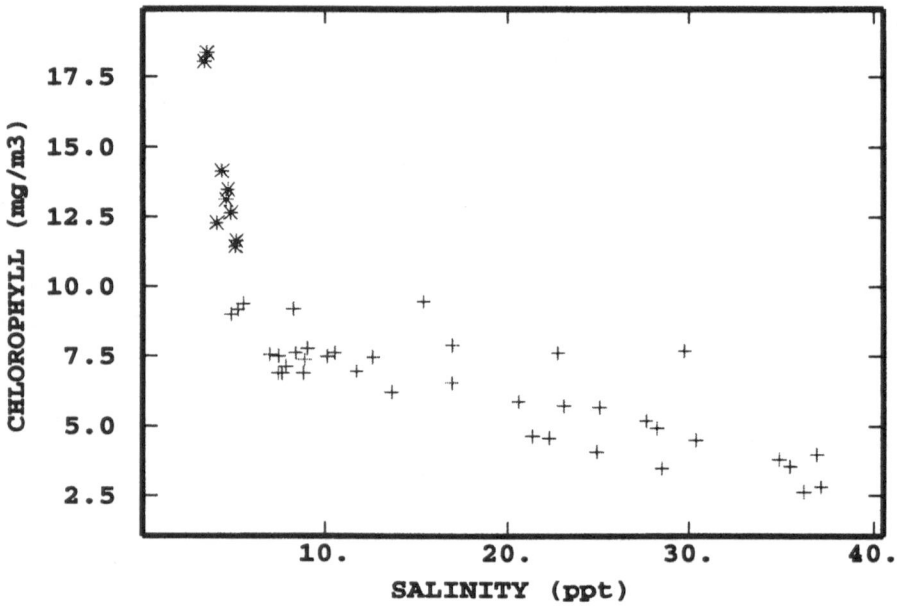

Fig. 3.7. Scatter diagram between salinity and chlorophyll for the 47 water samples. Nine samples located in freshwater are plotted as stars, the others as crosses

Direct and cross variograms were computed for the three variables and were fitted using a nugget effect and a cubic model with a geometric anisotropy, taking a maximal range of 17km along the horizontal and a minimal range of 7.5m in the vertical. The fitting was done using an improved version of the algorithm of Goulard & Voltz (1992), running 200 iterations, restraining the fitting to distances less than 3km in the horizontal and 3m in the vertical, and taking weights proportional to the number of pairs in each direction, divided also by the average distances. The set of fitted direct and cross variograms is shown on Figure 3.8. The variogram of conductivity is fitted with a more generous nugget effect and a lower sill than it was by hand on Figure 3.4. The

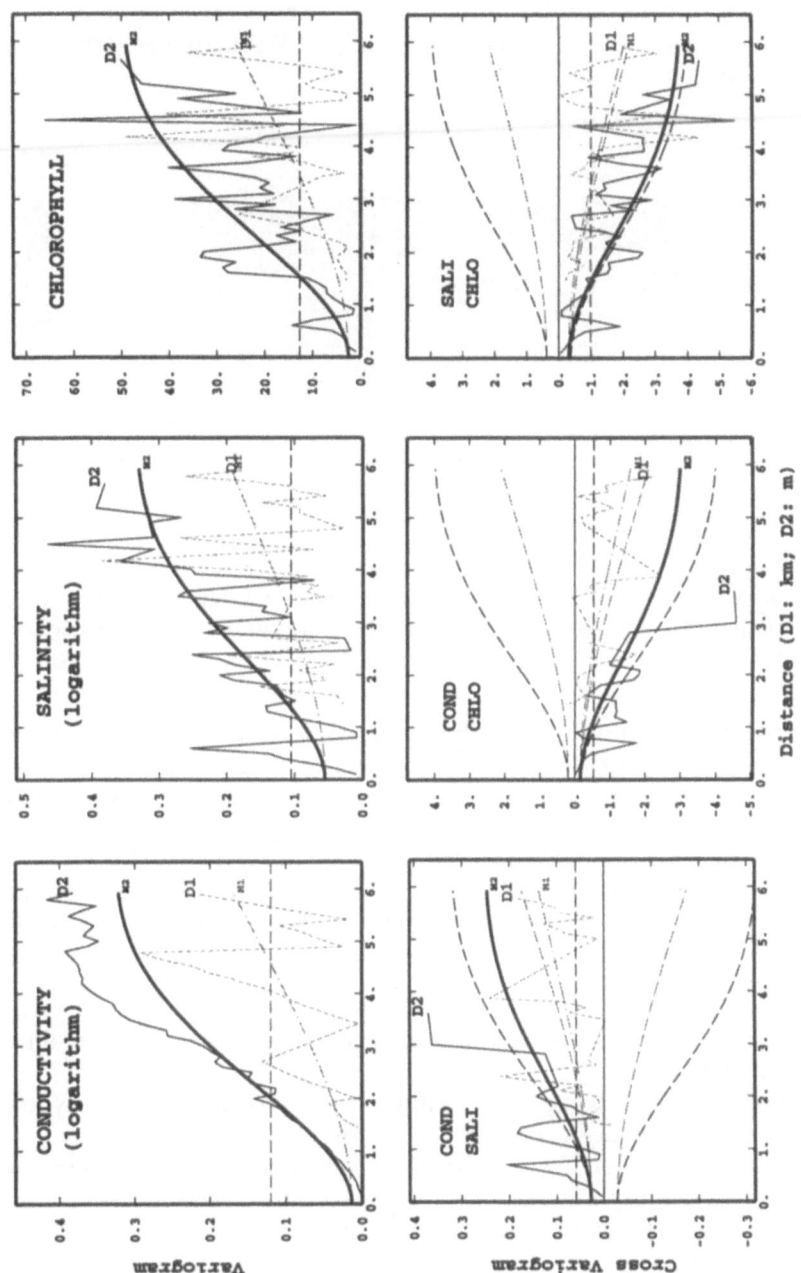

Fig. 3.8. Matrix of direct and cross variograms in two directions. The cross variograms additionally represent the hulls of correlation in both directions

cross variograms are displayed together with the correlation hulls computed from the corresponding direct variograms (see Wackernagel 1998 for further explanation).

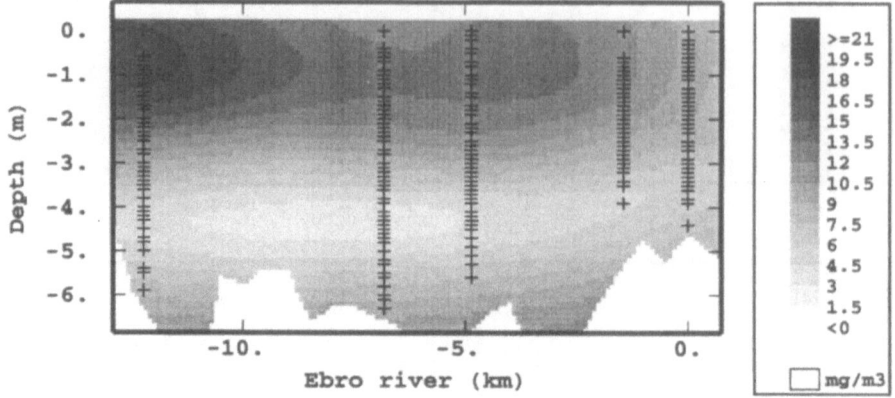

Fig. 3.9. Map of chlorophyll obtained by ordinary cokriging, filtering the nugget effect component, and including the logarithms of salinity and conductivity as covariates

The cokriging of chlorophyll taking logarithms of salinity and conductivity as auxiliary variables is displayed on Figure 3.9. The nugget effect, viewed as measurement error, was filtered. The use of conductivity as an auxiliary variable permits extrapolation of chlorophyll, quite successfully, at depths greater than where it was measured.
Extrapolation of chlorophyll in greater depth than where water samples are available depends much on how the model is formulated. Three different cokrigings using a nugget effect plus cubic model with ranges as defined above were experimented:

A the ordinary cokriging of chlorophyll with (untransformed) salinity and conductivity;
B the ordinary cokriging of chlorophyll with logarithms of salinity and conductivity (reference case described in detail above);
C the universal cokriging of chlorophyll with logarithms of salinity and conductivity, adding a linear drift in the vertical direction, to take account explicitly of the vertical non stationarity.

The three cokrigings essentially differ in the extrapolation behavior below a depth of 5m. The scatter diagrams of cokrigings A and C against B are plotted on Figure 3.10. The cokriging A (i.e. without taking logarithms, and thus neglecting the non linear relation with the auxiliary variables) grossly extrapolates at depths below 5m. This generates the cloud of values for which

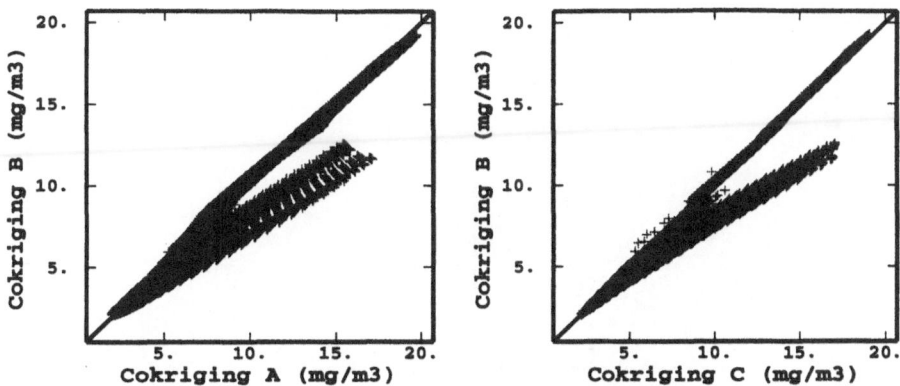

Fig. 3.10. Scatter diagrams of chlorophyll cokriging experiments A and C against B. The axes are in the same scale and the first bisector is drawn

cokriging A differs much from cokriging B. The cokriging C, adding a linear drift in the vertical, also yields higher values than case B in areas at greater depth, far from the water sample profiles.

3.6.3 Simulation of Chlorophyll

The variogram models of chlorophyll and salinity shown on Figure 3.8 cannot decently be qualified as 'fitting' the corresponding experimental variograms, because the latter exhibit little structure due to the small amount of data. So let us discuss this problem using the 47 chlorophyll data.

The experimental variogram of chlorophyll is shown on Figure 3.11 together with two models differing in their behavior at the origin:

- a geometrically anisotropic *cubic* model with a sill of 60 $(mg/m^3)^2$, with ranges 17km in the horizontal and 7.5m in the vertical.
- a geometrically anisotropic *exponential* model with a sill of 30 $(mg/m^3)^2$, with ranges 17km in the horizontal and 7.5m in the vertical.

A grid set at an origin (-12.9km,-5.2m) with 137 × 59 nodes using 100m × 10cm cells was defined for interpolation and stochastic simulation. Ordinary kriging was performed with variogram models and is shown on Figure 3.12 (the greytone scale used is the same as on Figure 3.9). The upper map represents the kriging with the cubic model while the lower map was obtained with the exponential model. We note that the cubic model with such a large range as compared to the dimensions of the domain has an extrapolative behavior, generating a maximum at -3km which is not supported by data.

The exponential model is more conservative: the maxima and minima in the map obviously refer to the highest and lowest values in the chlorophyll data. However this has also another implication: the kriged value will be more like the kriged mean of the domain the more we move away from data locations.

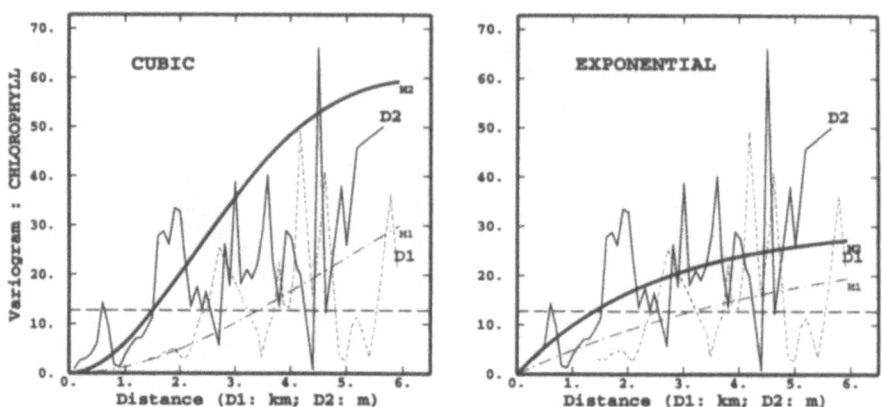

Fig. 3.11. The variogram of chlorophyll has been "fitted" with two models having a different behavior at the origin

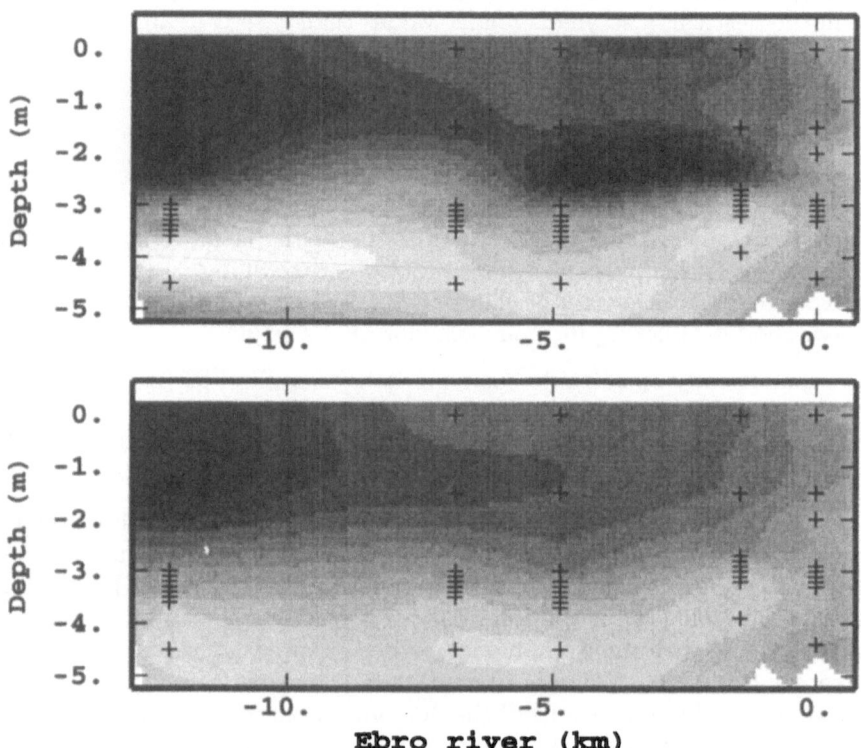

Fig. 3.12. Ordinary kriging of chlorophyll using the cubic (upper map) and the exponential (lower map) variogram models

It is well known that kriging does not represent an attempt to reconstruct the regionalized variable (had we the ability to measure it everywhere in the domain). The regionalized variable in the geostatistical model is but one realization of the random function. Kriging can be thought of as the average of many realizations that coincide with the data at sample locations. Thus if we are interested in what the regionalized variable at hand might look like, we have to employ *conditional simulation* instead of kriging.

We use the turning bands method for simulating realizations of the random function. We shall assume that the 47 chlorophyll values can be considered as a few samples of a realization of a Gaussian random function as the histogram does not indicate an asymmetric distribution. One thousand bands were used for simulation – with only 100 bands some bands could be seen with naked eye on the simulated map of the exponential model. The latter is no surprise when the range of the variogram is larger than the size of the domain. The simulation with a thousand bands is no big deal on modern computers.

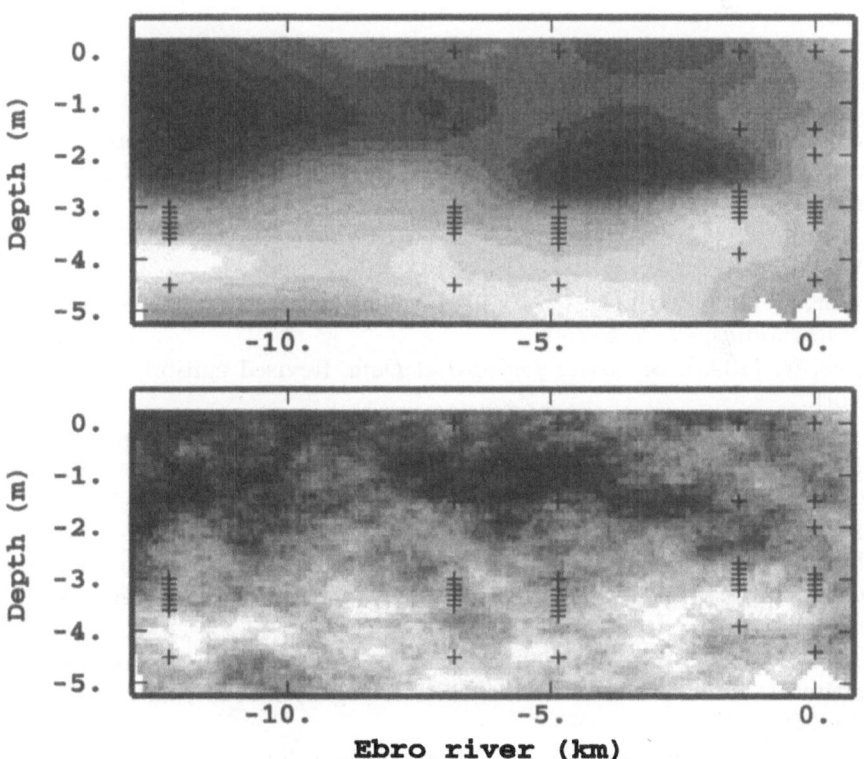

Fig. 3.13. Conditional simulations of chlororophyll with the turning bands method using the cubic (upper map) and the exponential (lower map) variogram models

Two conditional simulations are shown on Figure 3.13 (the greytone scale used is the same as on Figure 3.9). The upper map illustrates the fact that a surface corresponding to the realization of a random function with a cubic variogram is smooth, which is related to the parabolic shape of this model at the origin. The lower map shows that the realization of a random function with an exponential variogram has a rough aspect, due to the non differentiability of that model at the origin.

In applications the smoothness/roughness of the regionalized variable, when it is known, can be an important criterion for selecting the type of variogram model to be employed when the data, like in the present case, are not sufficient to characterize properly the behavior at the origin of the variogram.

Acknowledgments

This work was carried out in the framework of the European Community research project MAS3-CT98-0170 PIONEER. Computations were performed with the version 3.3 of the software Isatis. We wish to thank Xavier Emery and Jacques Rivoirard for helpful comments.

References

Chica-Olmo, M. & Abarca-Hernandez, F. (1998). Radiometric coregionalization of Landsat TM and SPOT HRV images. *International Journal of Remote Sensing*, **19**, 997–1005.

Chilès J.P., Delfiner P. (1999). *Geostatistics: Modeling Spatial Uncertainty*. Wiley, New York.

Cramér, H. (1940). On the theory of stationary random processes. *Annals of Mathematics*, **41**, 215–230.

Cressie, N. (1993). *Statistics for Spatial Data*. Revised edition, Wiley, New York.

Gomez-Hernandez, J., Soares, A. & Froidevaux, R., eds. (1999). *geoENV II – Geostatistics for Environmental Applications*. Kluwer Academic Publishers, Dordrecht.

González del Río, J., Falco, S., Sierra, J. P., Rodilla, M., Sánchez-Arcilla, A., Romero, I., Rodrigo, J., Martínez, R., Benedito, V., Aparisi, F., Mösso, C. & Movellán, E. (2000). Nutrient behaviour in Ebro river estuary. Submitted.

Goulard, M. & Voltz, M. (1992) Linear coregionalization model: tools for estimation and choice of multivariate variograms. *Mathematical Geology*, **24**, 269–286.

Grzebyk, M. & Wackernagel, H. (1994). Multivariate analysis and spatial-temporal scales: real and complex models. *Proceedings of XVIIth International Biometrics Conference*, Volume 1, 19–33, Hamilton, Ontario.

Hudson, G. & Wackernagel, H. (1994). Mapping temperature using kriging with external drift: theory and an example from Scotland. *International Journal of Climatology*, **14**, 77–91.

Lindner, S. & Wackernagel, H. (1993). Statistische Definition eines Lateritpanzer-Index für SPOT/Landsat-Bilder durch Redundanzanalyse mit bodengeochemischen Daten. In: Peschel G (ed) *Beiträge zur Mathematischen Geologie und Geoinformatik*, Bd 5, 69–73, Sven-von-Loga Verlag, Köln.

Matheron, G. (1965). *Les Variables Régionalisées et leur Estimation*. Masson, Paris.

Matheron, G. (1982). Pour une Analyse Krigeante des Données Régionalisées. Technical Note N-732, Centre de Géostatistique, Fontainebleau, France, 22p.

Monestiez, P., Allard, D., Navarro-Sanchez, I. & Courault, I. (1999). Kriging with categorical external drift: use of thematic maps in spatial prediction and application to local climate interpolation for agriculture. In: Gomez-Hernandez, J., Soares, A. & Froidevaux, R., eds.(1999).   *geoENV II – Geostatistics for Environmental Applications*. Kluwer Academic Publishers, Dordrecht, 163–174.

Myers, D.E. (1991). Pseudo-cross variograms, positive-definiteness, and cokriging. *Mathematical Geology*, **23**, 805–816.

Rivoirard, J. (2000). Which models for collocated cokriging? *Mathematical Geology*, **33** (2).

Sierra, J.P., González del Río, J., Sánchez-Arcilla, A., Movellán, E., Rodilla, M., Mösso, C., Martínez, R., Falco, S., Romero, I. & Marotta, L. (2000a). Dynamics of the Ebro river estuary and plume in the Mediterranean Sea. Proceedings of 10th International Biennial Conference on Physics of Estuaries and Coastal Seas, Norfolk, Virginia, USA, in press.

Sierra, J.P., González del Río, J., Sánchez-Arcilla, A., Flos, J., Movellán, E., Rodilla, M., Mösso, C., Falco, S., Romero, I. & Cruzado, A. (2000b). Spatial distribution of nutrients in the Ebro estuary and plume. Submitted.

Vargas-Guzmán, J. A. & Yeh, T. C. J. (1999). Sequential kriging and cokriging: two powerful geostatistical approaches. *Stochastic Environmental Research and Risk Assessment*, **13**, 416–435.

Wackernagel, H. (1998). *Multivariate Geostatistics: An Introduction with Applications*. 2nd ed., Springer-Verlag, Berlin.

Xu, W., Tran, T. T., Srivastava, R. M. & Journel, A. G. (1992). Integrating seismic data in reservoir modeling: the collocated cokriging alternative. *Proceedings of 67th Annual Technical Conference of the Society of Petroleum Engineers*, paper SPE 24742, 833–842, Washington.

Part II

Environmental Sampling and Standards

4 Distance Sampling: Recent Advances and Future Directions

S. T. Buckland, L. Thomas, F. F. C. Marques, S. Strindberg,
S. L. Hedley, J. H. Pollard, D. L. Borchers, and M. L. Burt

University of St. Andrews,
School of Mathematics and Statistics,
North Haugh, St Andrews KY16 9SS, Scotland

Abstract. We briefly describe distance sampling, then review recent advances that enhance its usefulness for wildlife and conservation management:

- Covariate models for the detection function.
- Double-platform methods for when detection on the line or at the point is uncertain.
- Spatial distance sampling models.
- Indirect distance sampling surveys.
- Automated design algorithms.
- Adaptive distance sampling designs.

The standard software for distance sampling analyses is described, and plans for future versions are outlined. In many distance sampling surveys, the main goal is to estimate trends in abundance over time. We anticipate that wildlife managers will in future wish to embed population dynamics models in these trend estimates, and we outline a state-space framework that allows this.

4.1 Introduction

Distance sampling is the most widely used technique for estimating abundance of wild animal populations. The most popular distance sampling method is line transect sampling, in which an observer travels along lines, laid down according to some randomized design, and records detected animals, and their distance from the line. In the closely-related point transect method, an observer visits a number of points, and records all animals detected from each point, together with their distances from the point.

We review these techniques briefly; a comprehensive treatment is given by Buckland *et al.* (1993). We also describe advances that enhance their usefulness for wildlife and conservation management.

Covariate models for the detection function potentially yield more efficient estimates of abundance, and eliminate the bias that arises for example when there is size- biased detection of animal clusters. They also offer the potential of more reliable estimates of trend in abundance when surveys are conducted from platforms of opportunity, such as ferries or fishing vessels at sea, for which the manager cannot randomly sample the survey region.

One of the key assumptions in distance sampling is that any animal on the line or at the point is detected. For some marine species such as whales or porpoise, or burrowing animals such as rabbits or tortoise, this assumption is violated. In 'double-platform' methods, distance sampling methodology is combined with mark-recapture, to allow estimation for such species. The second 'platform' may be a standard sightings platform on a ship or aircraft, or it may comprise an independent method of locating a subset of the surveyed animals, such as a radio-tagging experiment.

Spatial line transect modelling allows a surface representing animal density throughout the study region to be fitted. By integrating over the relevant section of the surface, abundance for any subset of the area may be estimated. Spatial models also relate abundance to spatial covariates, so that managers can assess the importance of habitat and environment to the population of interest.

Indirect distance sampling surveys widen the applicability of distance sampling methods by sampling not the animal itself but the sign it produces. Usually this is dung (e.g. elephants, deer, large cats) or nests (e.g. apes). The estimated density of sign must then be converted to an estimate of animal density.

Automated design algorithms, linked with GIS functionality, enable different designs to be compared for efficiency and accuracy of the subsequent abundance estimates, using simulation. For complex surveys in which coverage probability is not uniform, they also allow estimation of coverage probability by location. This in turn allows valid abundance estimation, using the Horvitz-Thompson estimator. (Conventional distance sampling methods assume that the coverage probability is uniform, at least within a stratum.)

For populations that typically have an aggregated spatial distribution, adaptive distance sampling surveys potentially yield more precise estimates of abundance than conventional surveys. They also give more detections with the same overall effort, which can be an important advantage for scarce species, for which sample size may otherwise be inadequate for modelling the detection function.

The above advances are currently being implemented in version 4 of the software DISTANCE. DISTANCE Version 3.5 (Thomas *et al.* 1998) was released in late 1998, and is the standard software for analysis of distance sampling data.

Generally, wildlife managers and conservation biologists are interested not just in population size but also trends in size over time. Trends are usually estimated empirically. We show how it is possible to estimate trends with an embedded population dynamics model. This allows more reliable prediction under different management scenarios than is possible with purely empirical modelling, so that managers can explore the implications of different management practices with greater confidence. We show how the models can be fitted using state-space methods.

4.2   Standard Distance Sampling Methods

4.2.1   Quadrat Sampling

Distance sampling methods are an extension of quadrat sampling. In quadrat sampling, quadrats are laid down at random and the number of animals (or more usually plants) of interest within each quadrat is counted. If k quadrats, each of area a, yield a total count of n animals, then animal density is estimated as

$$\hat{D} = \frac{n}{ka} \tag{4.1}$$

Extensions to two special cases of quadrat sampling lead to the two main distance sampling techniques. The first special case is strip transect sampling, in which quadrat i is a long, narrow rectangle of length l_i and width $2w$ (w either side of the centreline). Denoting total length of the rectangles by $L = \sum_i l_i$, (4.1) becomes

$$\hat{D} = \frac{n}{2wL} \tag{4.2}$$

The other special case of interest is point counts, in which the k quadrats are circles each of radius w, so that $a = \pi w^2$, and (4.1) becomes

$$\hat{D} = \frac{n}{k\pi w^2} \tag{4.3}$$

4.2.2   Line Transect Sampling

In line transect sampling, an observer travels along a line, and records the perpendicular distance x of each detected animal from the line. Sometimes,

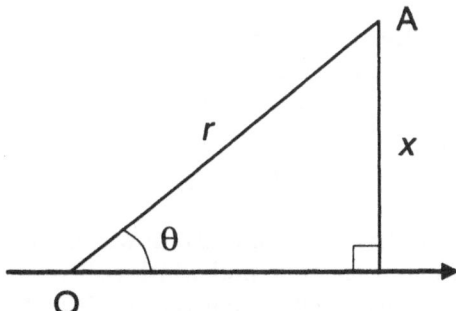

Fig. 4.1. An animal A is detected while the observer is at O. The perpendicular distance x is often estimated by estimating the detection distance r and sighting angle θ, from which $x = r\sin(\theta)$.

it is easier to record the detection or radial distance r and angle θ (Figure 4.1), from which $x = r\sin(\theta)$. The design generally comprises k lines, or a systematic grid of lines, randomly placed in the survey region. The random placement ensures that animals are uniformly distributed with respect to their distances from the line. It is assumed that all animals on or very close to the line are detected, but probability of detection may decrease with distance from the line, out to some distance w (which might be infinite). A

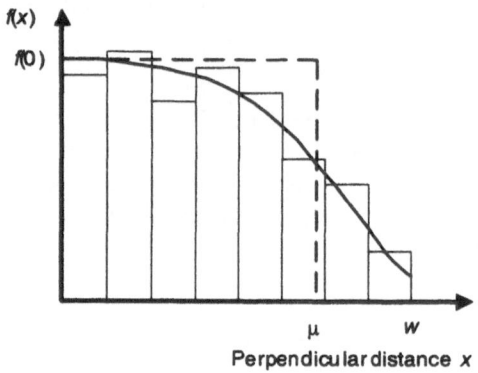

Perpendicular distance x

Fig. 4.2. Histogram of perpendicular distances x and the associated probability density function $f(x)$. The effective half-width μ is the distance such that the expected number of animals detected beyond μ (represented by the area under $f(x)$ for $\mu \le x \le w$) is equal to the expected number missed within μ (represented by the area within the rectangle of width μ and height $f(0)$ that lies above $f(x)$). Note that, because these two areas are equal, and $\int_0^w f(x)\,dx = 1$, then the area of the rectangle of width μ and height $f(0)$ is also one, so that $\mu = 1/f(0)$.

probability density function (pdf) $f(x)$ is fitted to the distance data, which in turn allows the effective half-width μ of the surveyed strip to be estimated (Figure 4.2): μ is the distance such that we expect to detect as many animals beyond μ as we fail to detect within μ of the line. Animal density is then estimated as

$$\hat{D} = \frac{n}{2\hat{\mu}L} \qquad (4.4)$$

This is clearly a simple extension of (4.2), in which the strip half-width w is replaced by the estimated effective half-width $\hat{\mu}$.

When animals occur in well-defined clusters, such as flocks or schools, then the cluster becomes the recorded unit for line transects (but not for strip transects). If n_s clusters are detected, and $\hat{E}(s)$ denotes the estimated mean size of clusters in the population, then cluster density is estimated by (4.4), with n_s replacing n, and estimated animal density is:

$$\hat{D} = \frac{n_s \hat{E}(s)}{2\hat{\mu}L} \qquad (4.5)$$

4.2.3 Point Transect Sampling

A point transect survey in its simplest form comprises k points randomly positioned throughout the study region, or more usually a systematic grid of points randomly superimposed on the region. The observer visits each point, and records the detection distance r from the point to each detected animal.

All animals on or very close to the point are assumed to be detected, and

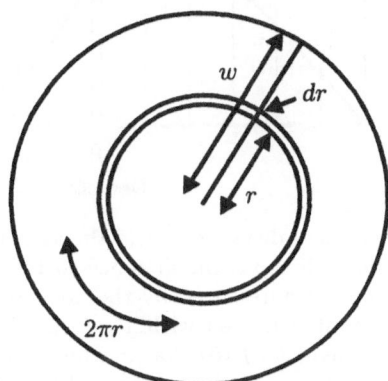

Fig. 4.3. The annulus of width dr at a distance of r from the point has approximate area $2\pi r\, dr$, and an animal within the annulus is detected with probability $g(r)$. It follows that $f(r)\, dr \propto 2\pi r g(r)\, dr$, where the constant of proportionality is $1/\int_0^w 2\pi r g(r)\, dr$, to ensure the honesty condition.

again, probability of detection is assumed to decrease with distance from the point, out to distance w. In line transect sampling, the pdf of perpendicular distances $f(x)$ is identical in shape to the probability of detection, expressed as a function of distance from the line ($g(x)$). By contrast, in point transect sampling, the area of an incremental annulus of radius r increases linearly with r (Figure 4.3), so that estimation of the effective radius ρ from the fitted pdf $\hat{f}(r)$ (Figure 4.4) is a little more complicated. Having obtained $\hat{\rho}$, animal density is estimated as

$$\hat{D} = \frac{n}{k\pi\hat{\rho}^2} \tag{4.6}$$

where w from (4.3) has been replaced by $\hat{\rho}$.

Equation (4.6) is modified in the same way as was (4.4) when animals occur in clusters.

4.3 Covariate Models for the Detection Function

Suitable properties of a model for the detection function $g(y)$, where $y = x$ is the perpendicular distance in line transect sampling or $y = r$ in point transect sampling, are:

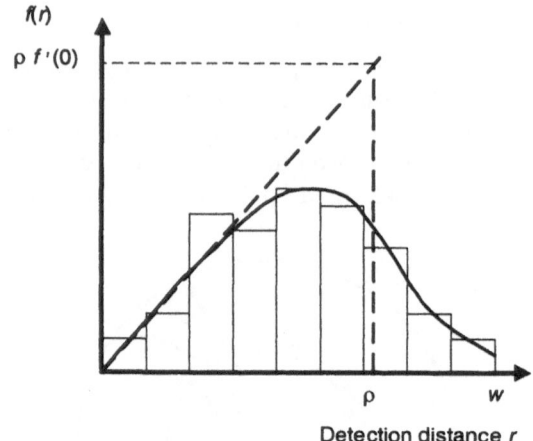

Fig. 4.4. Histogram of detection distances r and the associated probability density function $f(r)$. The effective radius ρ is the distance such that the expected number of animals detected beyond ρ (represented by the area under $f(r)$ for $\rho \le r \le w$) is equal to the expected number missed within ρ (represented by the area within the triangle of width ρ and height $\rho f'(0)$ that lies above $f(r)$). Note that, because these two areas are equal, and $\int_0^w f(r)\,dr = 1$, then the area of the triangle of width ρ and height $\rho f'(0)$ is also one, so that $\rho = \sqrt{2/f'(0)}$.

1. $g(0) = 1$;
2. the function has a 'shoulder': $g'(0) = 0$;
3. $g(y)$ is non-increasing over the range $[0, w]$;
4. $g(y)$ has the flexibility to fit a wide range of shapes.

Buckland *et al.* (1993) used a parametric key with the first three properties listed above, then allowed polynomial or cosine series adjustments to the fit of that key, to meet the fourth requirement. The number of adjustments can be selected for example using AIC or likelihood ratio tests. The keys they used were the half-normal ($g(y) = \exp[-y^2/2\sigma^2]$), the uniform ($g(y) = 1/w$), and a model derived from hazard-rate modelling of the detection process ($g(y) = 1 - \exp[-(y/\sigma)^{-b}]$). For the uniform, the scale 'parameter' is the truncation distance w. However, for the other two keys, the scale parameter σ is estimated using maximum likelihood. This means that σ itself can be a function of covariates, and conditional on those covariates, the likelihood can be maximized with respect to all parameters (including the covariate coefficients) in the usual way. Ramsey *et al.* (1987) were the first to use this approach. Marques and Buckland (in preparation) use it to 'fit covariate models for the detection function, using the half-normal or hazard-rate keys together with polynomial or cosine adjustments. The software has been implemented in DISTANCE 4 (Thomas *et al.* in preparation).

For populations that do not occur in clusters, $f(0)$ (line transects) or $f'(0)$ (point transects) may be replaced by the mean value across detected animals, from which μ or ρ may be estimated (Marques and Buckland, in preparation). However, for clustered populations, cluster size is likely to be one of the covariates, in which case (4.5), in conjunction with this approach, is biased. This is due to size-bias: larger clusters will tend to be over-represented in the sample of detected clusters, so that if estimation of $E(s)$ is not integrated with the covariate modelling, $\hat{E}(s)$ will be biased upwards. We can avoid this problem by using a Horvitz-Thompson estimator in which the inclusion probabilities have been estimated. For line transects, this yields:

$$\hat{D} = \frac{\sum_{i=1}^{n} s_i / \hat{\mu}(z_i)}{2L} = \frac{\sum_{i=1}^{n} s_i \hat{f}(0|z_i)}{2L} \tag{4.7}$$

where s_i is the size of the ith detected cluster and z_i is the vector of covariates for that cluster (which may include its size s_i) (Borchers et al. 1998a; Marques and Buckland, in preparation).

Covariate modelling can be useful for improving precision of abundance estimates and for accommodating the effects of size-biased sampling in estimation. They also allow researchers to gain understanding in how different variables affect detectability (e.g. Figure 4.5). Many marine surveys are conducted from 'platforms of opportunity', where there are insufficient funds to allow a suitable survey vessel to be hired, but adequate funds to place observers on board vessels that are passing through the region of interest for another purpose. Covariate methods have the potential to reduce the bias from such non-random surveys, and improve the reliability of estimated trends in abundance (Marques and Buckland, in preparation).

4.4 Double-Platform Methods

In some surveys, detection on the trackline is not certain ($g(0) < 1$), perhaps because some animals are underground or under water, or simply hidden by vegetation, when the observer passes. In this case, mark-recapture methods may be combined with distance sampling (Borchers, Zucchini and Fewster 1998b). In the case of marine mammal surveys, this is achieved by having two observation platforms. These might be treated as mutually independent, so that, provided that animals detected by both platforms ('duplicate detections') can be identified, two-sample mark-recapture methods that incorporate covariates (Huggins 1989) can be used. Bias in such methods is typically large unless heterogeneity in detectability is well-modelled. However, it is seldom possible to record covariates that reflect this heterogeneity adequately; for example if a whale produces a blow that is particularly visible from one platform, due to light conditions or some other factor in the environment that is difficult to measure, then it will tend to be more visible from the other platform too, and abundance will be underestimated. These problems

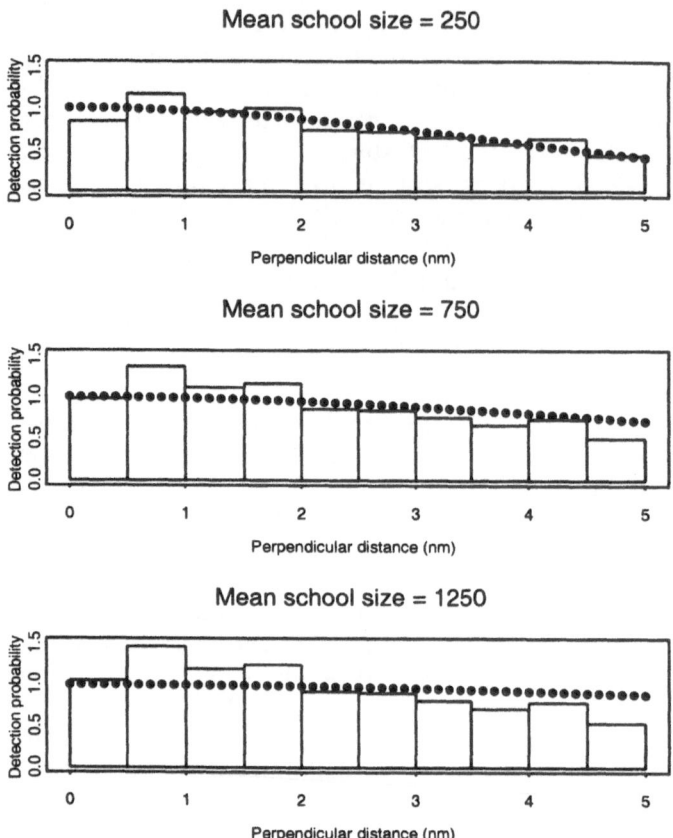

Fig. 4.5. Estimated detection functions for schools of the north-eastern offshore stock of spotted dolphin in the eastern tropical Pacific. Note the increase in detectability as school size increases.

may be reduced by separating the areas of search for the two platforms, and using one to set up trials for the other. Thus the 'tracking' platform detects animals, tracks them into the area searched by the 'primary' platform, and records whether they are detected by the primary platform. These binary data may then be estimated by logistic regression. Borchers *et al.* (1998a) used this approach to estimate harbour porpoise abundance in the North Sea. The average detection function from that example is shown in Figure 4.6.

In some surveys, the second 'platform' might be provided for example by a radio-tracking study, in which locations of tagged animals at the time of the sightings survey can be determined. These then provide the trials; we record

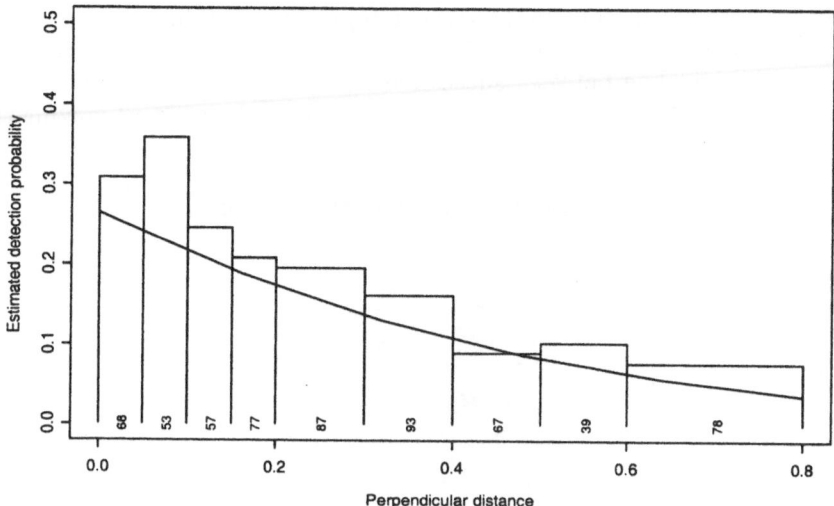

Fig. 4.6. Perpendicular distance distribution of detections made by the primary platform in surveys of harbour porpoise in the North Sea. Histogram bars show the proportion of tracked pods that are subsequently detected by the primary platform. Also shown is the average fitted logistic detection function, in which Beaufort sea state and stratum were modelled as factors, in addition to the continuous covariate, perpendicular distance. The fitted detection function appears low relative to the histogram bars due to the effect of modelled heterogeneity. Average detection probability on the trackline is estimated to be about 0.27.

which of the tagged animals are recorded in the distance sampling survey, and analyse the resulting data using logistic regression. To achieve adequate sample size, it may prove necessary to set up additional transect experiments, directing the observer past animals whose location is known (but not to the observer).

4.5 Spatial Distance Sampling Models

Spatial distance sampling models are potentially useful for several reasons: animal density may be related to habitat and environmental variables, improving understanding of factors influencing abundance; animal abundance may be estimated for any sub-region of interest, simply by numerical integration of the estimated density surface over the sub-region; and they potentially allow analysis of data collected from transects that are non-random, such as in 'platforms of opportunity' surveys.

One way in which spatial models can be fitted to distance sampling data is to conceptualize the distribution of animals as an inhomogeneous Poisson process, in which the detection function represents a thinning process. If in the case of line transect sampling, the data are taken to be distances along the transect line between successive detections, this allows us to fit a spatial surface to these data. We can refine this further by conceptualizing the

Density of harbour porpoise schools

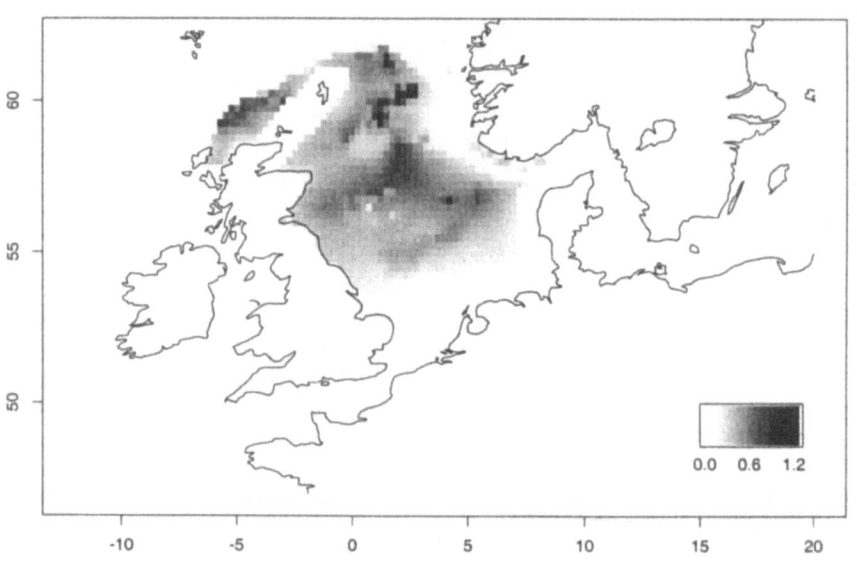

Fig. 4.7. Fitted density surface, harbour porpoise in the North Sea, 1994. High densities occur over the shallow banks of the North Sea, but the species is largely absent from the southernmost areas and from the English Channel, possibly as a result of high incidental catches in fishery operations.

observations as 'waiting areas', that is the effective area surveyed between one detection and the next, where the effective width of the surveyed strip varies according to environmental conditions and observer effort (Hedley, Buckland and Borchers 1999, in preparation). If the methods are applied to the example of harbour porpoise in the North Sea (see previous section), we obtain the spatial surface of Figure 4.7.

4.6   Indirect Distance Sampling Surveys

Sometimes it is more practical to survey not the animal itself but something it produces. This may be because the animal typically moves away from the observer before it can be detected (such as deer in dense forest), or because the animal is too rare, so that a large amount of effort would be required to yield a sufficient sample of direct sightings (such as large cats or apes). Usually, dung is surveyed (Marques *et al.* in press), although for apes, nests are surveyed (Plumptre 2000). Here, we assume that dung piles are being surveyed.

Density of dung piles P is estimated by a standard line transect survey (4.4):

$$\hat{P} = \frac{n}{2\hat{\mu}L} \qquad (4.8)$$

We now divide by an estimate of ξ, the mean number of days for a dung pile to decay, to obtain $\hat{G}$, the estimated number of dung piles produced per day per unit area:

$$\hat{G} = \frac{\hat{P}}{\hat{\xi}} \qquad (4.9)$$

By dividing this by $\hat{\eta}$, the estimated daily production of dung by one animal (number of dung piles per day), we obtain $\hat{D}$, our estimate of animal density:

$$\hat{D} = \frac{\hat{G}}{\hat{\eta}} = \frac{\hat{P}}{\hat{\xi}\hat{\eta}} \qquad (4.10)$$

The method is described more fully by Marques *et al.* (in press) and Buckland *et al.* (in preparation).

4.7   Automated Design Algorithms

Geographic Information Systems (GIS) are now widely available. To exploit them fully, designers of sightings surveys require automated design algorithms. This allows them to generate designs with known properties rapidly and simply. It also allows designs to be generated on the survey platform as they are required, which is an important advantage if the design must be able to accommodate rapidly changing circumstances. An example is the annual minke whale surveys of waters bordering the Antarctic ice edge; the ice can move over 100km in a day, so that the survey must be designed as it progresses. Another advantage of automated design algorithms is that they may be applied to simulations of the population of interest, allowing the user to assess the bias and precision in estimates obtained from each possible design. Design algorithms are usually straightforward. Most designs are stratified, in which case decisions that might be assessed by simulation include the number of strata, and where to locate the stratum boundaries. Within each stratum, points might be randomly located, or on a regular grid that is randomly superimposed on the stratum. If the stratum is large, a cluster of points might be surveyed at each sample location. For line transect sampling in which lines run right across the stratum, lines are typically chosen to be parallel. They might be randomly spaced, or systematically spaced with a random start, or some intermediate strategy, such as one line randomly located with each of k strips that cover the full stratum. If lines are short relative to stratum size, then design options are similar to those for point transects.

Sometimes, more complex algorithms are required. For example, shipboard surveys typically use continuous, zig-zag survey lines ('samplers'), so that costly ship time is not wasted in travelling from one line to the next. For convex survey regions or strata, a sampler with even coverage probability can be obtained by defining a principal axis for the design, and varying the angle of the sampler to that axis as the ship progresses through the area, to ensure that coverage probability is constant with respect to distance along the principal axis (Strindberg and Buckland, in preparation). By contrast, fixed-angle samplers do not give even coverage probability unless the survey region is rectangular. Fixed-waypoint samplers, which pass through equally spaced points on opposite sides of the survey region boundary, partially but not fully address this problem (Figures 4.8 and 4.9).

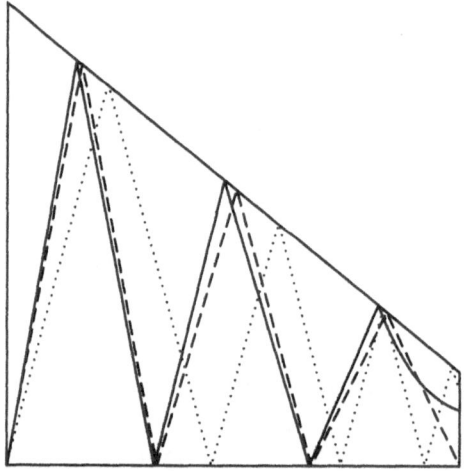

Fig. 4.8. A trapezoidal survey region illustrating three zig-zag samplers: equal-angle (dotted line); equal-waypoint (dashed line); and even-coverage (solid line). The principal axis of the design is parallel to the base of the trapezium in this example, and for the equal-waypoint sampler, waypoints are equally spaced with respect to distance along the principal axis, alternating between the top boundary and the bottom boundary (the base).

If the survey region or stratum is not convex, several options exist. First, if the region can be split into a small number of convex sections, an even-coverage sampler may be placed in each section, such that coverage probability is the same in each section. Second, a convex hull might be placed around the region, and an even-coverage sampler located within the hull. Sections of line outside

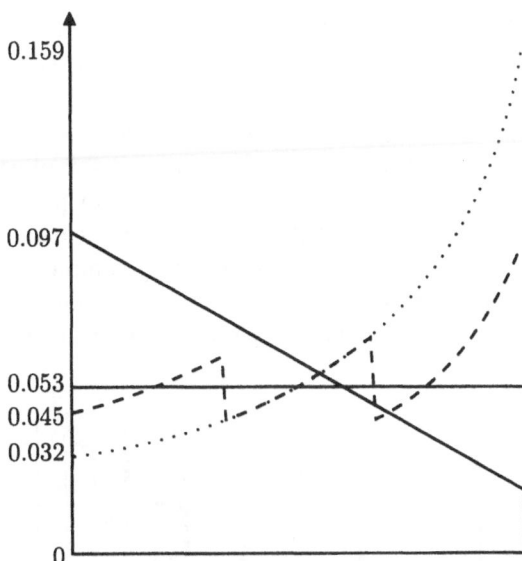

Fig. 4.9. Coverage probability against distance along the principal axis for the three designs of Figure 4.8. Also shown is the height of the trapezium as a function of distance along the principal axis, which indicates that the equal-angle sampler has too low coverage where the study region is wide, and too high where it is narrow. For the equal-waypoint sampler, coverage probability changes at each waypoint, and between waypoints, varies smoothly in the same manner as the equal-angle sampler.

the region would then not be covered. If neither of these options is practical, possibly due to complex coastal topography, a third option is to use an ad hoc algorithm, and evaluate coverage probability by location, using simulation. Abundance may then be estimated using Horvitz-Thompson to allow for the varying coverage probability. The properties that such an algorithm should possess include the following: no part of the survey region should have zero probability of cover; and probability of cover should be as nearly constant as possible.

4.8   Adaptive Distance Sampling

Animal populations are typically aggregated, and even large sightings surveys often yield small sample sizes of scarce species. These two features mean that abundance estimates are often imprecise, and the detection function $g(y)$ may be poorly modelled from the available data, perhaps leading to substantial bias. Adaptive sampling (Thompson and Seber, 1996) offers a means of increasing sample size, and hence improving both precision and bias, at no extra cost. Standard adaptive sampling methods can be extended to distance sampling. For point transect sampling, this is readily achieved by defining a grid of points, randomly superimposed on the study region, and randomly or systematically sampling from the grid for the primary sample. When additional effort is triggered at a sample point, points from the grid that neighbour the sample point are sampled. Coverage of these points may then trigger further sampling. Adaptive line transect sampling may be carried out in a similar way, but instead of defining a grid of points, we can define a grid of rectangles, and place a centreline within each rectangle. If this design yields lines that are too short, the primary sample can be taken to be a systematic or random sample of parallel strips of these rectangles, where each strip traverses the study region (Figure 4.10). Whenever coverage of a

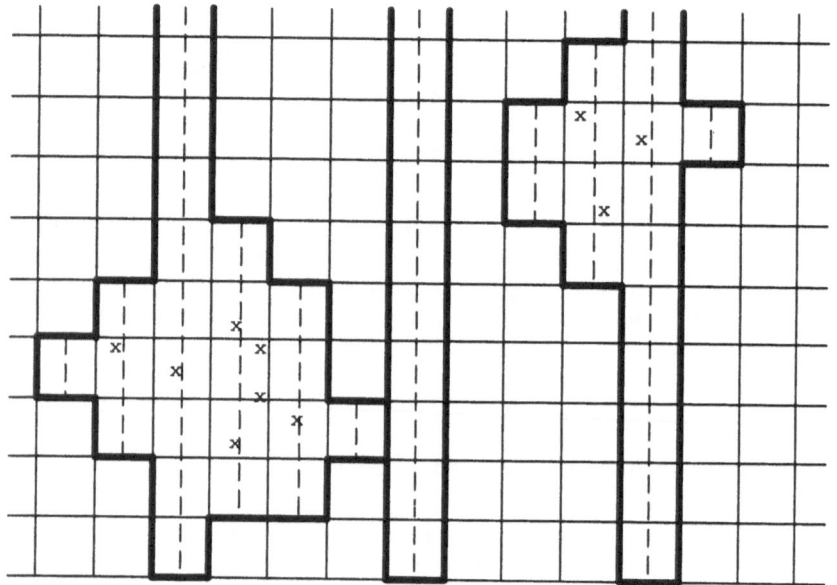

Fig. 4.10. An example of an adaptive line transect survey. Part of the study region is shown, overlaid by a grid. A systematic sample of columns from this grid comprises the sample of primary units. Additional effort is triggered whenever an animal, indicated by ×, is detected; the resulting surveyed areas are indicated by the thick lines, and the corresponding transects by dashed lines.

rectangle triggers additional effort, the neighbouring rectangles are sampled, which may then trigger sampling of their neighbours. In distance sampling, detection of animals within the surveyed strip or plot is not certain, so that the inclusion probabilities of standard adaptive sampling must be modified appropriately.

A major practical problem of adaptive sampling is that the required survey effort is not known in advance. This is particularly problematic for shipboard surveys, in which the ship is available for a predetermined number of days. Thus standard adaptive sampling will either finish early, so that ship time is wasted, or run out of effort, in which case the study region will be incompletely surveyed. Additionally for shipboard surveys, we would wish to avoid time off-effort, while the ship proceeds from one line to another; the loss of on-effort time may well more than offset any gain in efficiency from using adaptive sampling. Pollard and Buckland (1997) and Pollard, Palka and Buckland (in preparation) developed a method in which, when additional effort is triggered, the ship changes to a zig-zag (and hence continuous) course, centred on the nominal trackline (Figure 4.11). The angle of the zigzag is a

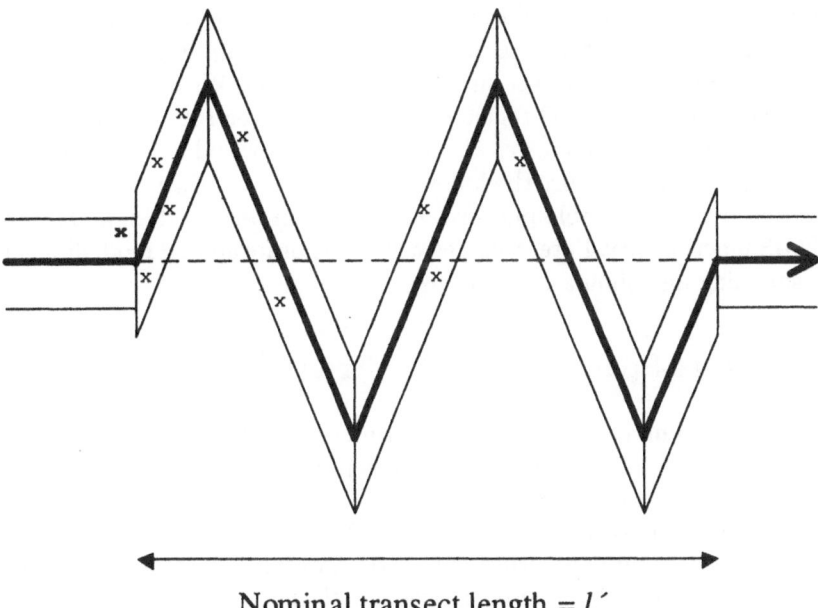

Nominal transect length $= l'$

Fig. 4.11. In a fixed-effort adaptive line transect survey, the amount of adaptive effort varies. In this example of a shipboard survey, the sighting of an animal (×) triggers the start of a zig-zag section of transect, in which the angle of the zig-zag depends on whether the survey is ahead of or behind schedule. If actual length of the zig-zag section of transect is l, and the nominal length is l', then data from the zig-zag section are given weight l'/l, relative to a weight of one for data from transects that follow the nominal route.

function of whether the ship is ahead of or behind schedule, allowing the total effort of the survey to be fixed in advance. This has the additional advantage that, if bad weather is encountered, the ship can make up for lost effort by reducing the amount of adaptive effort. In analysis, data from adaptive segments are downweighted to allow for the greater effort. Unlike standard adaptive sampling, the method is not design-unbiased, but simulations indicate that bias is small. The approach was tried experimentally on a survey of harbour porpoise in the Gulf of Maine. It yielded substantially more detections and significantly better precision than did conventional line transect sampling (Palka and Pollard, 1998; Pollard *et al.* in preparation).

4.9   The Software DISTANCE

The software DISTANCE was originally developed in parallel with the first book on distance sampling (Buckland *et al.* 1993). It replaced software TRANSECT, which had been written in conjunction with a monograph on line transect sampling (Burnham *et al.*, 1980). The first fully windows-based version of DISTANCE appeared in November 1998 (Thomas *et al.* 1998). Its capability was largely restricted to standard distance sampling; of the advances described above, only indirect surveys were included. Within two years of release, there were over 2000 registered users from over 100 countries. Version 4.0 of the software (Thomas *et al.* in preparation) was due for release in December 2000. It will feature the following advances: covariate modelling of the detection function; double-platform methods; automated design algorithms; and GIS functionality. Later releases will incorporate spatial modelling, adaptive sampling and simulation options.

4.10   State-Space Models for Trend

Distance sampling is widely used for estimating animal abundance. It is frequently the key tool for monitoring trends in abundance over time. The simplest method of estimating trend is simply to plot the abundance estimates against time. Often, it is useful to smooth these estimates, in an attempt to separate the sampling variation in estimates from their underlying trend. However, such smooths ignore knowledge of the population dynamics underlying changes in abundance. Thus, estimated trends may exhibit more rapid increase than is biologically possible for the species, or annual fluctuations may be larger than is plausible. In addition, this empirical approach to smoothing the trends does not allow us to project the trend into the future with any degree of confidence, or to examine the effects on these projections of different management strategies. These shortcomings may be alleviated to a large degree by embedding a population dynamics model within the trend analysis.

Fitting population dynamics models to a sequence of abundance estimates is not trivial, but three main approaches have been developed. Trenkel *et al.* (2000) used sequential importance sampling to fit a time series of deer count and cull data, Meyer and Millar (1999) and Millar and Meyer (2000) used MCMC to model fisheries data, and Newman (1998) used the Kalman filter to estimate mortality and movement of salmon. All three approaches could be used to fit a series of estimates of abundance from distance sampling surveys, although only the first two (which are both Bayesian) have the flexibility to cope with the non-linear, non-normal models that are often required to model population dynamics. The population dynamics model can be defined in the form of a state-space equation, and the population structure is related to the abundance estimates through an observation equation. For example, Trenkel *et al.* (2000) modelled deer populations using the state-space equation

$$n_t = S_{t-1}\left(R_{t-1}n_{t-1} - c_{t-1}\right) + e_t \tag{4.11}$$

where n_t is a vector whose elements represent number of animals in the population at time t in each age and sex class, the matrix R_t increments age of all animals by one and handles recruitment of young animals, the vector c_t subtracts animals harvested or known to have died in year t, the matrix S_t handles unobserved mortality, and the vector e_t represents stochastic errors. In their example, estimates of abundance of males aged ≥ 1, of females aged ≥ 1 and of calves were available, so the observation equation was

$$y_t = An_t + w_t \tag{4.12}$$

where the vector y_t comprises the three abundance estimates, A is an aggregation matrix that maps the population categories represented in vector n_t to the coarser categories for which abundance estimates are available, and the vector w_t represents stochastic errors.

4.11   Discussion

Distance sampling is the most widely used technique for assessing wildlife abundance. We have reviewed developments that (a) widen the range of species amenable to distance sampling surveys (species for which detection on the trackline is uncertain, and species that may be monitored more effectively by surveying their dung or nests); (b) increase the utility of the methods to wildlife managers (spatial modelling of animal abundance, assessment of factors that affect detectability and abundance, and the ability to fit a population dynamics model to a time series of abundance estimates); (c) improve the ease with which methods can be applied (automated design algorithms, and the windows-based software DISTANCE); and (d) improve the efficiency of the approach (adaptive distance sampling, covariate modelling, spatial modelling).

Distance sampling has attracted the attention of few statistical researchers relative to mark-recapture, despite its wider applicability. In the case of species for which both approaches are applicable, it is generally the more cost-effective option, and yields more reliable estimates of abundance, but it does not allow direct estimation of survival and recruitment rates. With the advent of Bayesian methods that allow us to fit population dynamics models to a time series of abundance estimates, this omission is addressed, although in the absence of other inputs, estimates of biological parameters will be imprecise. Mark-recapture might then offer invaluable input for specifying informative prior distributions on survival rates for example, which in turn allows other biological parameters to be estimated with greater precision. Similarly, any other knowledge of biological parameters can be incorporated through specifying informative prior distributions on them.

Fisheries assessment has traditionally been carried out by process modellers, reflecting the fact that the most readily available source of data has been catch data, and this needs to be related to a stock's population dynamics to allow reliable inference on stock size. Most other wildlife assessment work has been based on empirical modelling; even mark-recapture for open populations has generally failed to incorporate a biologically plausible population dynamics model, in which for example survival or recruitment might be density-dependent. The approach of fitting population dynamics models to series of population abundance estimates brings together the empirical modelling tools of statisticians with the process models employed by modellers.

References

Borchers, D.L., Buckland, S.T., Goedhart, P.W., Clarke, E.D. & Hedley, S.L. (1998a). Horvitz-Thompson estimators for double-platform line transect surveys. *Biometrics* **54**, 1221- 1237.

Borchers, D.L., Zucchini, W. & Fewster, R.M. (1998b). Mark-recapture models for line transect surveys. *Biometrics* **54**, 1207–20.

Buckland, S.T., Anderson, D.R., Burnham, K.P. & Laake, J.L. (1993). *Distance Sampling: Estimating Abundance of Biological Populations*. Chapman and Hall, London. 446pp.

Buckland, S.T., Anderson, D.R., Burnham, K.P., Laake, J.L., Borchers, D.L. & Thomas, L. in preparation. *Introduction to Distance Sampling*. Oxford University Press, Oxford.

Buckland, S.T., Goudie, I.B.J. & Borchers, D.L. (2000). Wildlife population assessment: past developments and future directions. *Biometrics*, **56**, 1–12.

Burnham, K.P., Anderson, D.R. & Laake, J.L. (1980). Estimation of density from line transect sampling of biological populations. *Wildlife Monographs* **72**.

Hedley, S.L., Buckland, S.T. & Borchers, D.L. (1999). Spatial modelling from line transect data. *Journal of Cetacean Research and Management* **1**, 255–64.

Hedley, S.L., Buckland, S.T. & Borchers, D.L., in preparation. Spatial models for line transect sampling.

Huggins, R.M. (1989). On the statistical analysis of capture experiments. *Biometrika* **76**, 133–40.

Marques, F.F.C. & Buckland, S.T., in preparation. Incorporating covariates into standard line transect analyses.

Marques, F.F.C., Buckland, S.T., Goffin, D., Dixon, C.E., Borchers, D.L., Mayle, B.A. & Peace, A.J. (In press). Estimating deer abundance from line transect surveys of dung: Sika deer in southern Scotland. *Journal of Applied Ecology*.

Meyer, R. & Millar, R.B. (1999). Bayesian stock assessment using a state-space implementation of the delay difference model. *Canadian Journal of Fisheries and Aquatic Sciences* **56**, 37-52.

Millar, R.B. & Meyer, R. (2000). Non-linear state space modelling of fisheries biomass dynamics by using Metropolis-Hastings within-Gibbs sampling. *Applied Statistics* **49**, 327–42.

Newman, K.B. (1998). State-space modeling of animal movement and mortality with application to salmon. *Biometrics* **54**, 1290–314.

Palka, D. & Pollard, J. (1999). Adaptive line transect survey for harbor porpoises. Pp. 3–11 in *Marine Mammal Survey and Assessment Methods* (eds Garner, G.W., Amstrup, S.C., Laake, J.L., Manly, B.F.J., McDonald, L.L. & Robertson, D. G.). Balkema, Rotterdam.

Plumptre, A.J. (2000). Monitoring mammal populations with line transect techniques in African forests. *Journal of Applied Ecology* **37**, 356–68.

Pollard, J.H. & Buckland, S.T. (1997). A strategy for adaptive sampling in shipboard line transect surveys. *Rep. Int. Whal. Commn.* **47**, 921-931.

Pollard, J.H., Palka, D. & Buckland, S.T. in preparation. Adaptive line transect sampling.

Ramsey, F.L., Wildman, V.W. & Engbring, J. (1987). Covariate adjustments to effective area in variable-area wildlife surveys. *Biometrics* **43**, 1-11.

Strindberg, S. & Buckland, S.T. in preparation. Optimised survey design: the zigzag sampler in line transect surveys.

Thomas, L., Laake, J.L., Derry, J.F., Buckland, S.T., Borchers, D.L., Anderson, D.R., Burnham, K.P., Strindberg, S., Hedley, S.L., Burt, M.L., Marques, F.F.C., Pollard, J.H. & Fewster, R.M. (1998). DISTANCE *3.5*. Research Unit for Wildlife Population Assessment, University of St. Andrews, UK. http://www.ruwpa.st-and.ac.uk/distance/

Thomas, L., Laake, J.L., Strindberg, S., Marques, F.F.C., Borchers, D.L., Buckland, S.T., Anderson, D.R., Burnham, K.P., Hedley, S.L. & Pollard, J.H. in preparation. DISTANCE *4.0*. Research Unit for Wildlife Population Assessment, University of St. Andrews, UK. http://www.ruwpa.st-and.ac.uk/distance/

Thompson, S.K. & Seber, G.A.F. (1996). *Adaptive Sampling*. Wiley, New York.

Trenkel, V.M., Elston, D.A. & Buckland, S.T. (2000). Calibrating population dynamics models to count and cull data using sequential importance sampling. *J. Amer. Statist. Assoc.* **95**, 363-374.

5 Setting Environmental Standards: A Statistical Approach

Vic Barnett and Marion Bown

The Nottingham Trent University,
Burton Street, Nottingham NG1 4BU, UK

Abstract. The EU, governments and agencies throughout the world seek to regulate and protect the state of the environment by setting environmental standards to control the level of certain pollutants present in various media of concern. It has become evident in recent years that many such standards are set without due consideration of uncertainty and variation and many are based on poorly defined principles. Barnett and O'Hagan (1997) conducted a study of current standards, concluding that most standards can be classified into one of two unsatisfactory categories; *ideal* and *realisable standards*. They proposed a new conjoint and more statistically meaningful concept; the *statistically verifiable ideal standard* (SVIS). We consider the implications of this approach with the aim of applying the results to standards formulation for practical problems of pollution in air, water and soil, working in co-operation with relevant bodies. We will discuss the case of a simple pollutant distribution, developing SVISs from various standpoints within a hypothesis-testing framework. We consider a *best linear unbiased quantile estimator* (BLUQE) for use in the compliance assessment of a SVIS and describe an approximate significance testing procedure using simulated critical values. Finally, we develop a SVIS for a practical example involving water quality and illustrate the use of the BLUQE.

5.1 Background

In recent years there has been much concern about the environment, leading to increased research activity in the investigation of environmental pollutants in the air, soil and water, and their effects upon health, ecosystems, climate and so on. In order to monitor and protect (and possibly improve) the current state of the environment, *environmental regulations* or *standards* are set to control the levels of such pollutants.

Those responsible for setting environmental standards are part of a hierarchical chain of liability, which can be broadly divided into three levels. Firstly, at the international level, informal standards may be derived through agreement between nations or may be enforced as formal international standards (such as those published by ISO). At the next level, national bodies such as the US Environmental Protection Agency or the UK Environment Agency impose standards to be implemented and monitored nation-wide, possibly encouraging individual states or local authorities to develop their own controls. Finally there are guidelines for self-regulation, where industries and businesses might set their own codes of practice. At each level, standards imposed must be at least as strict as their counterparts at higher levels, e.g. local standards must conform with national requirements.

In order to preserve or improve the well-being (potentially the health) of a subject group vulnerable to the pollution product, most environmental standards are set, or have a direct effect, upon the pollution source, and have to be complied with by those responsible for causing the pollution. Usually, but not exclusively, standards consist of some *upper* limit which pollutant levels should not exceed; otherwise the polluter may be penalised. Environmental standard-setting procedures have recently been investigated and critically evaluated in terms of the extent to which they take account of uncertainty and variability through the use of appropriate statistical methods. Barnett and O'Hagan (1997) explicitly examine how (if at all) prevailing standard-setting procedures account for uncertainty and variation, drawing a fundamental distinction between two types of current standards and proposing a new type of standard for future use. The book arose from evidence the authors prepared for the UK Royal Commission on Environmental Pollution and the new type of standard (to be discussed below) is endorsed in the Commission's report *Setting Environmental Standards* (RCEP, 1998).

Consider a situation where we have uncertainty about some pollution variable, X say. On one hand, we have a *population* of possible values of X characterised by a probability distribution, and on the other hand, we might draw a *sample* of observations from this population, from which we could make inference about the distribution. The vital problem with current standard-setting procedures is that frequently this inference process is missing, leading to two distinct classes of standard termed **ideal standards** and **realisable standards** (Barnett and O'Hagan, 1997).

Ideal standards consist of a statement about the general situation, i.e. the *population*, but do not indicate any practical statistical methodology by which compliance with the standard may be tested or monitored. An example is the 1997 UK National Air Quality Strategy standard for sulphur dioxide pollution in air, where the '15 minute mean' is not to exceed 100 p.p.b. We must assume that this means that the 15 minute mean is some parameter of the distribution of pollution values, and that this parameter must not exceed 100 p.p.b., but potential compliers are not told what is needed for them to demonstrate compliance.

In contrast, *realisable standards* state that the standard is met if some specified set of sample outcomes is obtained from a sampling procedure. However, this provides no formal statistical inference about the underlying population. For example, El-Shaarawi and Marsalek (1999) draw attention to water quality guidelines in Canada, part of which require that the geometric mean of the indicator bacteria *E.coli*, taken from at least 5 samples of water collected over a period of less than 30 days, should not exceed 2000 *E.coli* per litre. If any site is found to be in compliance with this standard, it merely implies that the *sample* taken is acceptable, but gives no formal information about the population of (the distribution of) pollution values over time or space. Neither of these current standard-setting procedures can be acceptable; we need to use the standard statistical machinery of model, sample and inference. This dissatisfaction prompted the more appropriate conjoint approach proposed by Barnett and O'Hagan (1997) in the form of the **statistically verifiable ideal standard** (a **SVIS** for short). The SVIS essentially links an ideal standard with a compliance criterion that must provide some prescribed level of statistical assurance that the ideal standard has actually been met. The linked statistical assurance requirement is the backbone of the SVIS approach. As an illustration, a SVIS might require that;

> *the 0.9 quantile of the pollutant distribution must not exceed level L, to be demonstrated with 95% confidence.*

Thus the standard becomes a conjoint statement of a required characteristic of the population of pollution values and of the statistical assurance that has to be provided that such a characteristic is being met. Some current standards do attempt to specify both an ideal standard and an operational compliance criterion (as in, for example, the US ozone standard from 1979 to 1997, as detailed in USEPA, 1997), but the two parts are never formally linked as in the SVIS and we do not know if meeting (failing) the compliance requirement ensures that the ideal standard is (is not) met.

Before we continue discussion of the SVIS, we should consider another important concept in environmental standard-setting; the **pollutant cause-effect chain.** Standards may be set at any point along this chain, which can be simplified to four steps. The first stage might include standards set upon the actions that actually *cause* pollution, e.g. production processes within factories. The second stage comprises standards set on pollutants at the *point of entry* to some medium (such as air, water or soil), and this includes the control of emissions from industrial chimneys and from cars. Thirdly and most often, standards are set to control the level of pollutant in the medium at the *point of contact* with the vulnerable subject group; for example, most air pollution standards are set at the level of ambient air. The final stage includes those standards set upon the *effect* of the pollution on the subject group; water quality guidelines in New Zealand provide a nice example, where the 'maximum acceptable' number of people developing swimming-associated illnesses is restricted to 19 per 1000 bathers (New Zealand Ministry for the Environment, 1998).

Establishing such a chain naturally leads to the following crucial question: how compatible are any separate standards set at different stages of the cause-effect chain? As an example, a standard that has been met on the cause of the pollution may turn out to be stricter than, or even inconsistent with, a subsequent standard set upon the pollutant at the point of contact with a subject group (further down the chain). So which, if any, standard should be adopted, or even adapted? To investigate such matters would require accurate modelling of the *pollutant cause-effect relationship*. This topic, in the spirit of *dose-response modelling*, and to which there are many statistical and deterministic approaches, would require major discussion beyond our aims here - for further details see for example Barnett and O'Hagan (1997), Barnett and Bown (2001a) and Risk Assessment Forum (1995).

5.2   Setting a Statistically Verifiable Ideal Standard

So how might developments in environmental standard-setting continue? Let us examine how a SVIS might be set on a pollutant at the point of contact with the subject group. Consider as an example the river water quality objectives set by the UK Environment Agency. These are realisable standards 'which samples of water are required to satisfy', which specify minimum sample sizes and sampling time windows (NRA, 1994). These standards are set upon *sample quantiles*.

However, we will examine a set of SVISs based upon *population quantiles* which could replace their current system. We could require a SVIS to be set upon a specific quantile (say the 0.99 quantile) of the distribution of the pollutant random variable X. For simplicity and illustration, we might assume a normal (or lognormal) distribution for X. A possible SVIS could take the following form:

> The 0.99 quantile, $\xi_{0.99}$, of the distribution of X should not exceed x_0.
> If this is true, then there should be no more than a 5% probability of
> misclassification as failure (non-compliance).

We could set this up in a hypothesis-testing framework within which we would identify what are effectively an *ideal standard* and a *compliance criterion*. So for the above SVIS, we could express the *ideal standard* component as a hypothesis

$$H_0 : \xi_{0.99} \leq x_0$$

to be contrasted with the natural alternative

$$H_1 : \xi_{0.99} > x_0.$$

The *compliance criterion* component would then be met, for example, by carrying out a significance test of H_0 against H_1 at level 0.05.

We might consider including a further requirement that fixes an additional degree of assurance of detecting true exceedance:

> Additionally, if $\xi_{0.99}$ were as high as $x_0 + k$ $(k > 0)$ i.e. a clear breach
> of the standard, this should be detected with probability of at least
> 0.95.

This means that our significance test above must have power at least 0.95 at $\xi_{0.99} = x_0 + k$ for a specified value of k. This would determine a minimum sample size needed for the test.

We feel that this is a reasonable and straightforward approach. But note that a major implication of this approach by the nature of a test of significance is that we are granting the *benefit of the doubt* to the potential polluter. That is, the null hypothesis of compliance will only be rejected if the sample quantile is sufficiently large. Thus the polluter is able to achieve a sample-estimated quantile that is possibly well in excess of x_0 before being declared as failing the standard.

By reversing the null and alternative hypotheses, we can assign the benefit of the doubt to the regulator, which is happening with some standards at present. An example includes the conservative 'divide-by-n' principle for compliance assessment using composite sampling, as currently employed by Australian contaminated site standards (Standards Australia, 1997); a more statistically meaningful standard for the situation is proposed by Barnett and Bown (2001b). But again this approach is problematical. Conscientious producers would then be forced by such a standard to limit pollution to levels much lower than x_0 to ensure that they are not assigned the default status of non-compliance.

Circumstances may justify one or other of these extreme standpoints; benefit of the doubt to the producer, or to the regulator. But in many cases, this will not be acceptable and we should aim to develop a more 'even-handed' approach. Current work is investigating this in terms of the setting of *two* standard levels, known as 'guard points', with certain requirements to be met at both points. This will be reported elsewhere.

5.3   A Best Linear Unbiased Quantile Estimator

Let us return to the SVIS discussed above. Operation of the test relies upon knowing the exact distribution of the pollutant levels. In most practical situations this is unlikely. Many data sets from pollution problems suggest positively skew distributions of pollution values, but often a normalising transformation is feasible. So it makes sense to consider the normal distribution, and we will develop the SVIS on this basis. Recall that the hypothesis test is set upon a population quantile i.e. $\xi_{0.99}$. For the normal distribution $N\left(\mu, \sigma^2\right)$ this takes a simple form $\xi_{0.99} = \mu + \zeta_{0.99}\sigma$ where $\zeta_{0.99}$ is the 0.99 quantile of the standard normal distribution. It is useful to seek to estimate this quantile directly in the case where the distribution parameters μ and σ are both unknown. We have chosen to approach this using best linear unbiased estimators (BLUEs) μ^* and σ^* for μ and σ, respectively (Hassanein *et al* (1986) also considered quantile estimation in this way). These estimators are calculated from linear combinations of the order statistics, and inevitably depend upon the means and variances of the reduced order statistics $U_{(i)} = (X_{(i)} - \mu)/\sigma$; further details may be found in David (1981).

Hence estimation of the quantile $\xi_{1-\gamma}$ can be carried out using the *best linear unbiased quantile estimator* (BLUQE):

$$Q^*_{1-\gamma} = \mu^* + \zeta_{1-\gamma}\sigma^*$$

where $\zeta_{1-\gamma}$ is the upper γ point of the standard normal distribution. To carry out the hypothesis test, we must compare $Q^*_{1-\gamma}$ with an appropriate critical value, which inevitably depends upon the sample size used to monitor compliance. In many compliance assessment problems, sample sizes are customarily small (for example, the UK Environment Agency monitors river pollutant levels *monthly*): up to 20 or so. Preliminary simulations with small random samples of size not exceeding 20 from a standard normal distribution showed that we could not adequately describe the distribution of $Q^*_{1-\gamma}$ as normal (or lognormal) in these cases. Since for the purpose of testing, only the high percentage points of the $Q^*_{1-\gamma}$ distribution interest us, we have conducted extensive simulations to calculate to reasonable accuracy the upper 5% and 1% points of the $Q^*_{1-\gamma}$ distributions with $\gamma = 0.05$ and 0.01, for samples of size not exceeding 20, and for the case of a standard normal parent distribution. These are presented in Table 5.1 below.

Table 5.1. Upper 5% and 1% points of the $Q^*_{1-\gamma}$ distribution with $\gamma = 0.05$ and 0.01, for $X \sim N(0, 1)$

n	$Q^*_{0.95} = \mu^* + 1.6449\sigma^*$		$Q^*_{0.99} = \mu^* + 2.3263\sigma^*$	
	$\alpha = 0.05$	$\alpha = 0.01$	$\alpha = 0.05$	$\alpha = 0.01$
2	4.27	5.63	5.88	7.74
3	3.47	4.35	4.74	5.94
4	3.12	3.81	4.25	5.18
5	2.92	3.50	3.98	4.75
6	2.78	3.30	3.80	4.47
7	2.69	3.15	3.67	4.27
8	2.61	3.04	3.56	4.12
9	2.55	2.95	3.48	4.00
10	2.50	2.87	3.41	3.89
11	2.45	2.81	3.36	3.81
12	2.41	2.75	3.31	3.74
13	2.38	2.70	3.26	3.67
14	2.35	2.66	3.22	3.62
15	2.32	2.62	3.19	3.57
16	2.30	2.59	3.16	3.53
17	2.28	2.55	3.13	3.49
18	2.26	2.53	3.11	3.45
19	2.24	2.50	3.09	3.42
20	2.23	2.48	3.07	3.39

But we need to consider $X \sim N(\mu, \sigma^2)$ where both μ and σ are unknown, for which we will estimate $\xi_{1-\gamma}$ by the BLUQE $Q^*_{1-\gamma}$. For a level-α significance test, we want to determine the critical value h such that

$$P\left(Q^*_{1-\gamma} > h \mid \xi_{1-\gamma} \leq x_0\right) \leq \alpha.$$

We can write the BLUEs μ^* and σ^* as

$$\mu^* = \sum_{i=1}^{n} \kappa_i X_{(i)} \quad \text{and} \quad \sigma^* = \sum_{i=1}^{n} \tau_i X_{(i)},$$

so that

$$P(Q^*_{1-\gamma} > h \mid \xi_{1-\gamma} \leq x_0) = P(\mu^* + \zeta_{1-\gamma}\sigma^* > h \mid \xi_{1-\gamma} \leq x_0)$$

$$= P(\sum \kappa_i X_{(i)} + \zeta_{1-\gamma} \sum \tau_i X_{(i)} > h \mid \xi_{1-\gamma} \leq x_0)$$

$$= P(\sum \kappa_i(\mu + \sigma U_{(i)}) + \zeta_{1-\gamma} \sum \tau_i (\mu + \sigma U_{(i)}) > h \mid \xi_{1-\gamma} \leq x_0)$$

$$= P\left(\sum \kappa_i U_{(i)} + \zeta_{1-\gamma} \sum \tau_i U_{(i)} > \frac{h - \mu(\sum \kappa_i + \zeta_{1-\gamma} \sum \tau_i)}{\sigma} \,\middle|\, \xi_{1-\gamma} \leq x_0\right)$$

$$= P\left(\mu_0^* + \zeta_{1-\gamma}\sigma_0^* > \frac{h - \mu}{\sigma} \,\middle|\, \xi_{1-\gamma} \leq x_0\right)$$

$$= P\left(Q^*_{1-\gamma}(0) > \frac{h - \mu}{\sigma} \,\middle|\, \xi_{1-\gamma} \leq x_0\right)$$

where $U_{(i)}$ is the reduced order statistic, $\sum \kappa_i = 1$ and $\sum \tau_i = 0$ (Gupta, 1952) and where $Q^*_{1-\gamma}(0)$ is the best linear unbiased $(1-\gamma)$ quantile estimator based on a random sample from $N(0, 1)$. But critical values for $Q^*_{1-\gamma}(0)$ have been tabulated above as a result of the simulation study. Hence if we let h_0 be the appropriate tabulated critical value given a particular combination of $\gamma \in [0.01, 0.05]$, $\alpha \in [0.01, 0.05]$ and $n \in \{2, \ldots, 20\}$, this implies that the critical value in the SVIS will be

$$h = \mu + h_0\, \sigma.$$

We may proceed to an approximate level-α test as follows. The null hypothesis

$$H_0 : \xi_{1-\gamma} = \mu + \zeta_{1-\gamma} \sigma \leq x_0$$

is rejected in favour of the natural alternative H_1 if

$$Q^*_{1-\gamma} > h,$$

that is, if

$$Q^*_{1-\gamma} > \mu + h_0\,\sigma$$
$$= \mu + \zeta_{1-\gamma}\sigma + (h_0 - \zeta_{1-\gamma})\,\sigma$$
$$= x_0 + (h_0 - \zeta_{1-\gamma})\,\sigma.$$

So we need to know the *excess* $(h_0 - \zeta_{1-\gamma})\sigma$; but this involves σ which is of course unknown. To seek an approximate test by replacing σ by the current BLUE σ^* is clearly unreasonable due to the inevitable high correlation between $Q^*_{1-\gamma} = \mu^* + \zeta_{1-\gamma}\sigma^*$ and the approximate critical value $h^* = x_0 + (h_0 - \zeta_{1-\gamma})\sigma^*$.

A more reasonable approach is the following. We propose that a matching second sample of size n be taken from which a further σ^* should be calculated; this estimate should then be substituted into the critical value calculation. Alternatively, we may have some knowledge of σ which we could use, or perhaps a value from an earlier study, for the calculation of the critical value h^*. Either method yields only an approximate test and the use of an external variance estimator may result in some loss of power (and perhaps a small change in the size) of the test. Simulations have been conducted to investigate this matter for the case $n = 12$, and suggest that the test size may be closer to 6% (rather than 5%). There is also modest loss of power in comparison with a test using internal statistics $\bar{X}$ and S in estimating the quantile; for example, a power loss of 6%, 18% and 3% at values for the quantile 25%, 50% and 100% above the hypothesized value.

5.4   Example: Copper Levels in River Water

To illustrate the testing approach, we consider an example based upon the UK river water quality objectives mentioned above. As part of the River Ecosystem Classification scheme, the Environment Agency (hereafter referred to as 'the Agency') must monitor the levels of copper and various other metals dissolved in the river water. Twelve monthly observations of copper levels are taken and the Agency's procedures require that we calculate 'the lower 95% confidence limit for the 0.95 sample quantile' (NRA, 1994). If this confidence limit exceeds the appropriate level, the river fails the standard for copper.

We could interpret the Agency's procedures by constructing a SVIS that employs a hypothesis test for the population quantile, instead of a sample value. Thus we would test

$$H_0 : \xi_{0.95} \le x_0 \quad \text{against} \quad H_1 : \xi_{0.95} > x_0$$

at the 5% significance level, where the standard level x_0 depends upon the hardness band of the river at this site. This test assigns the benefit of the doubt to the polluter and is analogous to the Agency's approach.

From a data set provided to the authors we consider a set of 12 observations of copper levels from the Long Buckby Wharf site on the Grand Union Canal from 1999 shown in Table 5.2 below (Environment Agency, 2000). Water in East Anglia is considered to be in the hardest band and so the appropriate standard limit for copper levels is $x_0 = 112\,\mu g/l$. We assume that copper levels follow a lognormal distribution (following Agency procedures) with unknown mean and standard deviation. Hence the corresponding standard limit on the transformed data is $x_0' = \ln(112) = 4.7185$. Calculation of the BLUQE estimator for the logged data set yields

$$Q_{0.95}^* = \mu^* + 1.6449\sigma^* = 2.1832 + 1.6449 \times 0.35695 = 2.7703.$$

Table 5.2. Copper observations at site R05BGGUCM250L on the Grand Union Canal, Northamptonshire, England: 1998 - 1999 (Environment Agency 2000)

Year	Copper Total, $\mu g/l$											
1998	32.6	10.2	7.4	9.2	5.4	11.5	11.9	10.1	16.4	9.9	8.7	6.7
1999	8.6	8.5	6.0	7.2	7.3	11.4	13.7	6.3	12.9	17.6	7.6	6.1

In this case, we do not have a second sample from which we can estimate σ; however, it seems reasonable to use an estimate obtained from the 1998 data at the same site. This yields the critical value

$$h = \ln(112) + (2.41 - 1.6449) \times 0.4623 = 5.0722,$$

where 2.41 is the appropriate tabulated value for a 0.95 quantile, tested at the 0.05 level of significance using a sample of size 12 (see Table 5.1). Since $2.7703 < 5.0722$, H_0 is not rejected at the 5% significance level and for this site, we would report that the standard for copper has been met.

5.5 Conclusion

We have tried above to illustrate how recent research into statistical environmental standard-setting procedures is developing from the proposals made in response to concern over the lack of acknowledgment of uncertainty and variation in current standards. Further work is extending these ideas in various directions based upon the notion of the SVIS which we, in agreement with the UK Royal Commission on Environmental Pollution (1998), feel should become central to the procedure of setting environmental standards. We hope that in future years, such research will lead to a well-needed revision of the many current environmental standards - a vast achievement for environmental statistical methodology.

Acknowledgments

Marion Bown is grateful to the EPSRC for their support through research grant 99801418. The authors are grateful also to the Environment Agency for providing the data set in Table 5.2.

References

Barnett, V. and Bown, M. H. (2001a). Standards, Environmental: Statistical Considerations. To appear in: El-Shaarawi, A. H. and Piegorsch, W. W. (eds) (2001). *Encyclopaedia of Environmetrics*. Wiley, New York.

Barnett, V. and Bown, M. H. (2001b). Statistically meaningful standards for contaminated sites using composite sampling. To appear in *Environmetrics*.

Barnett, V. and O'Hagan, A. (1997). *Setting Environmental Standards: the Statistical Approach to Handling Uncertainty and Variation*. Chapman & Hall, London.

David, H. A. (1981). *Order Statistics*. Second edn. Wiley, New York.

Department for the Environment, Transport and the Regions UK (DETR) (1999). *The Air Quality Strategy for England, Scotland, Wales and Northern Ireland*. HMSO, London.

El-Shaarawi, A. H. and Marsalek, J. (1999). Guidelines for indicator bacteria in waters: uncertainties in applications. *Environmetrics*, **10** (4), 521-529.

Environment Agency (2000). *Water Quality Data for the Grand Union Canal (Pollution Control Public Register)*. Personal communication.

Gupta, A. K. (1952). Estimation of the mean and standard deviation of a Normal population from a censored sample. *Biometrika*, **39**, 260-273.

Hassanein, K. M., Saleh, A. K. Md. E., and Brown, E. F., (1986). Best linear unbiased estimators for Normal distribution quantiles for sample sizes up to 20. *IEEE Transactions on Reliability*, **R-35** (3), 327-329.

National Rivers Authority (NRA) UK, 1994. *Water Quality Objectives: Procedures used by the National Rivers Authority for the purpose of the Surface Waters (River Ecosystem) (Classification) Regulations 1994*. NRA, Bristol.

New Zealand Ministry for the Environment (1998). *Recreational Water Quality Guidelines*. At: www.mfe.govt.nz/about/publications/water_quality/beaches-guidelines.htm

Risk Assessment Forum (1995). *The Use of the Benchmark Dose Approach in Health Risk Assessment*. EPA/630/R-94/007, US Environment Protection Agency, Washington, DC.

Royal Commission on Environmental Pollution (1998). *Setting Environmental Standards.* HMSO, London.

Standards Australia (1997). *Guide to the Sampling and Investigation of Potentially Contaminated Soil. Part 1: Non-Volatile and Semi-Volatile Compounds.* AS 4482.1-1997. Standards Australia, New South Wales, Australia.

US Environmental Protection Agency (USEPA) (1997). *National Ambient Air Quality Standards for Ozone; Final Rule.* At: `www.epa.gov/fedrgstr/EPA-AIR/1997/July/Day-18/a18580.htm`

Part III

Atmosphere and Ocean

6 The Interpretation and Validation of Measurements of the Ocean Wave Directional Spectrum

Lucy R. Wyatt

Sheffield Centre for Earth Observation Science, Department of Applied Mathematics, University of Sheffield, Sheffield S10 2TN, UK

Abstract. This paper discusses the problems that arise in interpreting and validating new technologies for the measurement of the directional spectrum of the ocean surface. Remote sensing techniques, from the ground or satellites, are under development but, in general, they require detailed numerical inversion methods to extract the wave information and hence need to be carefully validated. The more established measurement techniques, e.g. buoys, are themselves not perfect and do not measure the full spectrum, and these limitations needs to be taken into account in any validation work. The issues raised are discussed with reference to intercomparisons between HF radar and wave buoy measurements. By combining different intercomparison methods the limitations of the HF radar measurement are more clearly identified.

6.1 Introduction

Ocean surface waves, generated locally by the wind and often propagating long distances from their point of origin, play an important role in the transfer of momentum and particulates between the atmosphere and the oceans, in coastal erosion and associated sediment transports, and many other coastal and deep sea processes. Wave measurements and prediction are very important in fisheries and oil exploration and exploitation. A big issue in ship design at the present time is the impact of 'rogue' waves on ships of all sizes. The recent book (Junger 1998) and film 'The Perfect Storm' demonstrated this sort of wave and its impact. Wave forecasting is used in ship routing, i.e. determining the cheapest and safest route for ship transport. Harbour design and vessel traffic services, the authorities who control ship access into harbours, also rely on good wave measurements and forecasts. The economical assessment, siting and operation of wave power devices clearly need long term wave statistics as well as local forecasting. These are also important in the development of offshore wind power. In all these applications, knowledge about the direction of propagation of waves and the distribution of energy with wavelength, is becoming as important as information about their height. The wavelike response of the ocean surface to wind forcing can be demonstrated from the governing equations and boundary conditions. However the forcing and hence the response is not deterministic and any measurement or modelling requires a statistical description of the process. In this paper the impact of this on measurement and its validation is reviewed. More detailed discussions of these matters can be found in Krogstad *et al.* (1999) and Wyatt *et al.* (1999).

6.2   The Statistical Description of the Sea Surface

The ocean surface is usually thought of as a random surface composed of waves of different directions and wavelengths with random amplitudes and phases. This is generally described by the ocean wavenumber directional spectrum, $S(\mathbf{k})$, or directional frequency spectrum, $S(f, \theta)$. This describes the average distribution of energy with wavenumber $\mathbf{k}$ (or frequency f and direction θ). From this we can derive the frequency spectrum, $E(f) = \int_0^{2\pi} S(f, \theta)\, d\theta$, and Fourier coefficients of the directional distribution,

$$a_n(f) = \frac{\int_0^{2\pi} S(f, \theta) \cos\theta\, d\theta}{E(f)}$$

and

$$b_n(f) = \frac{\int_0^{2\pi} S(f, \theta) \sin\theta\, d\theta}{E(f)}.$$

From the Fourier coefficients the following directional parameters are derived: mean direction,

$$\theta_m(f) = \tan^{-1}\left(\frac{b_1(f)}{a_1(f)}\right)$$

and mean directional spread,

$$\sigma_m(f) = \left(2\left[1 - \sqrt{a_1(f)^2 + b_1(f)^2}\right]\right)^{\frac{1}{2}}.$$

Even at the parametrised level of $E(f)$, $\theta_m(f)$ and $\sigma_m(f)$ there is still a lot of information although most intercomparisons of different wave measurement techniques confine themselves to a few parameters that represent measures of the total energy. These are obtained by integrating over the measured frequency range (f_1 to f_2) which is often from about 0.03Hz to 0.5Hz. The parameters used are significant waveheight,

$$H_s = 4\sqrt{\int_{f_1}^{f_2} E(f)\, df},$$

mean (or first moment) period,

$$T = \frac{\int\limits_{f_1}^{f_2} E(f)\, df}{\int\limits_{f_1}^{f_2} f\, E(f)\, df},$$

mean direction,

$$\theta_{mean} = \tan^{-1} \frac{\int\limits_{f_1}^{f_2} \sin \theta_m(f) E(f)\, df}{\int\limits_{f_1}^{f_2} \cos \theta_m(f) E(f)\, df},$$

and mean spread,

$$\sigma_{mean} = \tan^{-1} \frac{\int\limits_{f_1}^{f_2} \sin \sigma_m(f) E(f)\, df}{\int\limits_{f_1}^{f_2} \cos \sigma_m(f) E(f)\, df}.$$

Significant waveheight can be shown to be roughly equal to the average height of the highest one third waves (for discussion of this point see Tucker 1991). It ranges from less than 1m in calm conditions to over 10m in storms in the open ocean. The second moment period is sometimes used instead of T. This is related to the mean zero up-crossing period but is a more unreliable measurement since it depends on the accuracy of high frequency measurements.

6.3   Measurement Techniques

The most commonly used wave measurement devices are either single point instruments or small arrays. Wavebuoys measure accelerations or displacements due to wave action and are now beginning to use GPS (Global Positioning System) technology. Arrays of capacitance wave gauges measure spatial differences in water level. Bottom-mounted sensors measure wave-induced pressure and current variations using a combination of pressure sensors and acoustic Doppler profilers. Relationships between these measurements and many wave parameters are reasonably well known but they do not provide direct measurements of the full directional spectrum. To obtain this, maximum entropy (Lygre & Krogstad 1986, Hashimoto 1997), maximum likelihood (Capon 1969) and wavelet (Donelan et al. 1996) methods are used.

In recent years remote sensing wave measurement techniques have been developing. HF (High Frequency: 3-30MHz) radar systems are located on the coast or on board ships or offshore platforms. These measure backscatter from the sea surface over ranges from 1 to more than 100km in many different directions with spatial resolution (depending on radar parameters) of $\sim$ 1 to 10 km^2. The directional spectrum at each location is measured by inverting a non-linear integral equation – a model which successfully describes backscatter in low to moderate seas (Wyatt 1995). Fig. 6.1 shows an example of H_s and wind direction (Wyatt *et al.* 1997) measured by the WERA HF radar during the EuroROSE Fedje (on the west coast of Norway) experiment. Most of the work on HF radar wave measurement has been carried out at

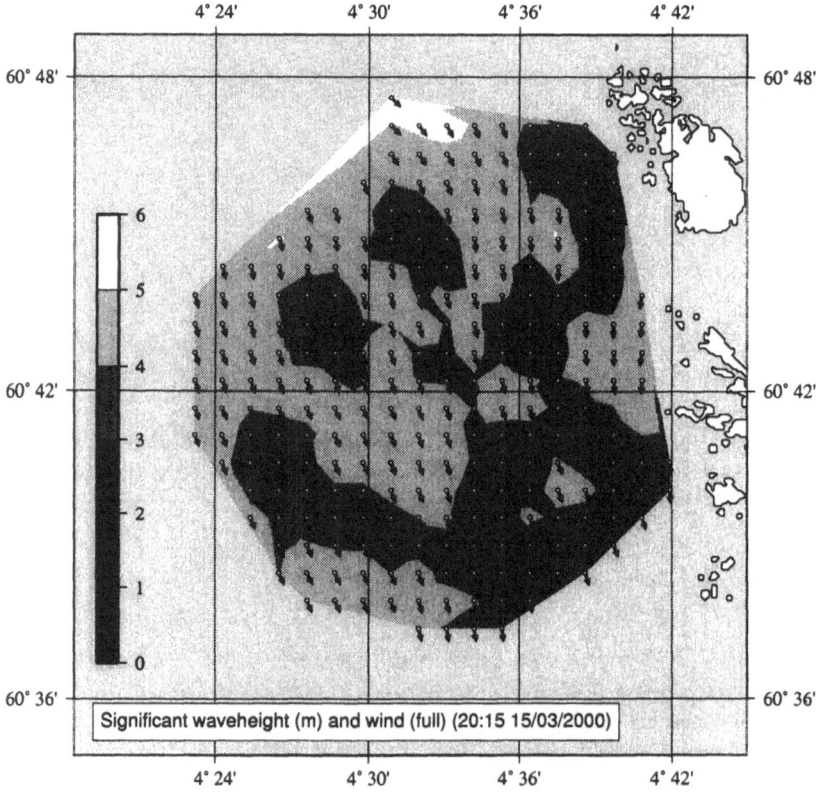

Fig. 6.1. Significant waveheight (in shaded grey scales) and wind direction (arrows) measured from the west coast of Norway during the EuroROSE Fedje experiment.

the University of Sheffield (Wyatt 1990, 1991, 1995, 2000, Atanga & Wyatt 1997, Wyatt *et al.* 1992, 1994, 1996, 1999). Examples from this work are used to illustrate points made in the rest of this paper. Comparisons with other instruments have been carried out over a wide range of conditions during experiments in the Southern North Sea (Wyatt et al. 1998), the Celtic Sea (Wyatt et al. 1986, Wyatt 1991), the North Atlantic and Bay of Biscay (Günther et al. 2000) and the East Coast of the US. Ground- (or ship-) based microwave radars operate at much higher frequencies and measure backscatter over much smaller ranges (Borge et al. 1999). The directional spectrum is measured using a 3-D Fourier transform of time varying images of the surface. Microwave radars mounted on aircraft or satellites are also being used to measure wave parameters. Satellites have the capability to provide measurements on global scales. Altimeters (which can also be mounted on offshore platforms) measure the time taken for a radar pulse to travel from the radar to the sea surface and back again and hence provide a waveheight measurement (Bauer & Heimbach 1999, Krogstad & Barstow 1999). Synthetic aperture radar generates a radar image of the sea surface from which the directional spectrum can be measured by inverting a complex integral equation (the backscatter model) assuming some prior information (Hasselman & Hasselman 1991, Krogstad & Barstow 1999).

6.4   Validation/Intercomparison Issues

New technologies need to be evaluated against old and accepted technologies. All measurements are subject to errors. These arise from a number of different sources, listed below, and all need to be taken into account in the evaluation process.

- Differences and inherent limitations in measurement principles (e.g. HF radar and wavebuoys have very different measurement principles – radio wave scattering versus surface accelerations).
- Systematic offsets associated with poor calibrations.
- Inherent and different sampling variability (e.g. wave buoys sample the ocean surface in time, whereas HF radar essentially samples in time and space).
- Temporal and/or spatial offsets (e.g. HF radars measure integrated backscatter over an area of, say, 1km^2 which is then compared with the point measurement of a buoy which may or may not lie exactly in the HF radar measurement region. The measurement time period may also be different.)
- Requirement for stationarity and/or homogeneity. In this discussion stationarity will be defined as constancy of statistics with respect to time and homogeneity with respect to space. Waves respond to weather systems which can change on time scales of minutes to hours to days so stationarity cannot be assured. Spatial inhomogeneities are introduced by variations in bottom depth, in currents and weather features such as fronts all of which could occur on spatial scales of less than 1km^2.

For now- and fore-casting purposes there is usually a requirement to assimilate the data into wave models. Error estimates are needed for these applications.

6.5   Methods Used and their Application

It is perhaps surprising that many published intercomparisons present only qualitative information with no additional statistics to quantify any judgements or conclusions made. Differences in the time and location or spatial coverage are often used to justify any observed differences in the measurements. It is however true that there is as yet no satisfactory quantitative test of the significance of the difference between two measurements of the full directional spectrum. Qualitative judgements are needed and are very useful, but only in combination with statistics of the comparison of derived parameters of the spectrum. As has already been said, the buoy measurements do not provide the full directional spectrum directly and the processing methods used with these instruments also require verification. Fig. 6.2 shows a comparison between an HF radar measured spectrum and one obtained from a buoy using maximum entropy analysis. There is qualitative agreement between the two but it is hard to distinguish, in a quantitative manner, differences due to inadequacies in the radar inversion from those associated with the maximum entropy processing.

In the following discussion the reference measurement will be referred to as x and the measurement to be validated as y. Where statistics are given they most commonly assume x is 'sea-truth' and hence attribute any differences to errors in y. Some of the following statistics are often used.

- Correlation coefficient (or T-linear correlation for circular functions)
- Mean error (bias) and standard deviation
- Scatter index (standard deviation/mean(x))
- Mean square error
- Mean relative (to x) error and standard deviation
- Regression coefficients

However x is usually a buoy measurement and the errors in this ought to be taken into account in the analysis. Methods used that do take this into account often assume equal and known (or estimated) variances in x and y (not necessarily constant) and then use relatively straightforward variations on standard regression. In the following discussion the word 'error' needs to be interpreted as 'difference'.

Our approach has been to estimate individual variances of x and y since there is no physical reason why they should be equal, particularly when we are dealing with remote sensing compared with buoy measurements. The problem then becomes one of estimating parameters of a functional relationship model using maximum likelihood. i.e we now model the reference and data as

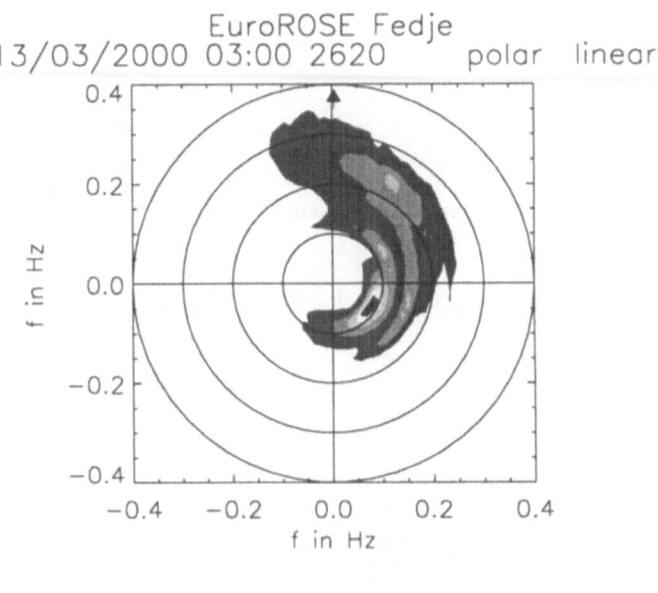

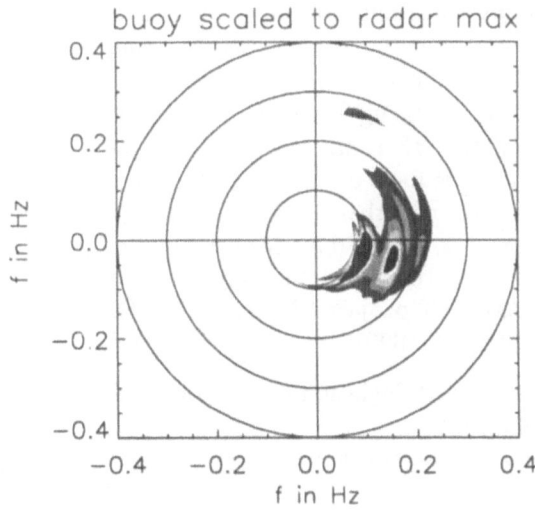

Fig. 6.2. Directional spectra measured by the radar (above) and wave buoy (below) during the EuroROSE Fedje experiment. Six grey scaled levels are shown with the darkest denoting the location of the peak. The arrow in the upper plot is the radar measured wind direction. Circles are drawn at frequencies of 0.1,0.2,0.3 and 0.4Hz. These plots show swell waves with a peak frequency of about 0.09Hz propagating from the north-west and a second system in roughly the same direction with a peak at about 0.16Hz. The energy in the wind direction is much lower in the second system.

$$x = X + \delta$$

$$y = Y + \varepsilon$$

and assume the underlying variables X and Y satisfy a functional relationship model

$$Y = F(X) \quad \text{or} \quad \log Y = F(\log X))$$

e.g.

$$Y = \beta X + \alpha$$

where α and β are unknown parameters to be determined. Assuming δ and ε are normally distributed with known variances, we can estimate α and β using a maximum likelihood method. This method has been used extensively in HF radar/buoy comparisons (Sova 1995, Krogstad et al. 1999, Wyatt et al. 1999).

The impact of sampling variability on buoy spectral estimates is well established theoretically and, with the assumption that this is the only source of error that we need to account for, we can estimate variances for the derived parameters (e.g. H_s, θ_{mean}) using Taylor series. To estimate variance due to sampling variability of the HF radar measured parameterised spectrum i.e. $E(f)$, $\theta_m(f)$ and $\sigma_m(f)$, we have used Monte Carlo simulations (Sova 1995). The Taylor series method can then be applied to estimate the variance of derived parameters exactly as was done for the buoy data.

The application of these methods to intercomparisons between HF radar and buoy data in the SCAWVEX project (Wyatt et al. 1998) led to the following conclusions (Krogstad et al. 1999):

- Mean error and standard deviation (for H_s or T) is not a useful measure of accuracy or inaccuracy.
- Relative error is a better statistic and gives a reasonable variance estimate.
- The best estimate of bias (for H_s, T and θ) is obtained using maximum likelihood.
- Non-parametric regression could be used to suggest application of a non-linear functional relationship.
- Restricting attention to H_s, and mean T and θ is not sufficient to assess the accuracy of a measurement of the directional spectrum.
- To understand the sources of differences between two (or more) measurement systems, it is necessary to look at time series of parameter comparisons, and not rely on individual measurements, and take into account details of the directional spectra comparisons.

These conclusions have been substantiated in more recent experiments during the EuroROSE project (Günther *et al.*, 2000). Figures 6.3 and 6.4 show scatter plots of significant waveheight measured at Petten, the Netherlands (SCAWVEX) and Fedje, Norway (EuroROSE) respectively. Up to four su-

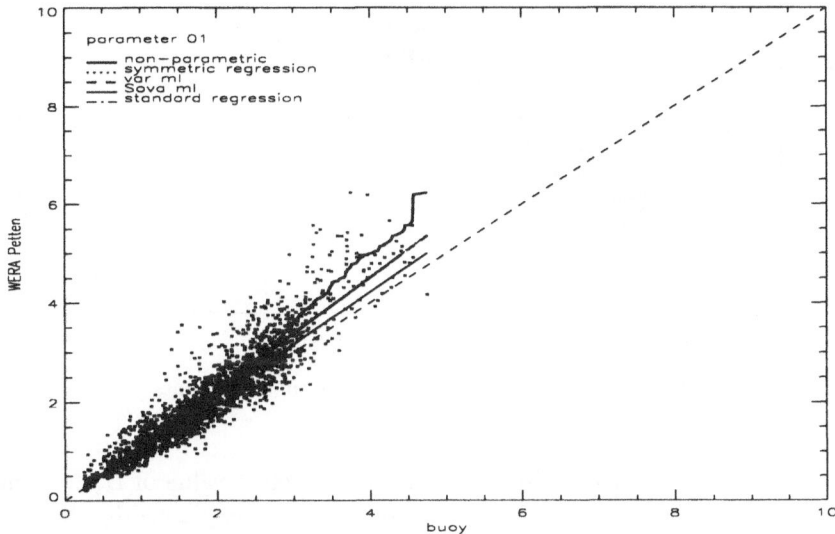

Fig. 6.3. Scatter plot of significant waveheight at Petten with buoy on the x-axis and radar on the y-axis. The thick solid line shows the non-parametric regression, dotted line is the symmetric regression, dash-dot the standard regression, dash the functional relationship with data determined variances and thin solid line the sampling variability functional relationship.

perimposed lines are seen on these plots. These are:

(a) the standard regression line (assuming no errors in the buoy data),
(b) a symmetric regression (assuming equal variances for buoy and radar),
(c) the maximum likelihood functional relation assuming Monte Carlo variance estimates for the radar and standard theory for the buoy and
(d) a maximum likelihood functional relation where the variances are estimated from the data (see below).

In the Petten case (a), (b) and (d) are very close, in the Fedje case (c) and (d) are close. In addition to these lines the non-parametric regression is shown with a solid line. In both cases all of these are difficult to distinguish below a certain value of H_s ($\sim$ 2m at Petten and $\sim$ 3m at Fedje). The non-parametric regression seems to be suggesting that the linear functional models are not appropriate over the full H_s range and this might explain the differences in slopes between the two experiments since the Fedje experiment has more

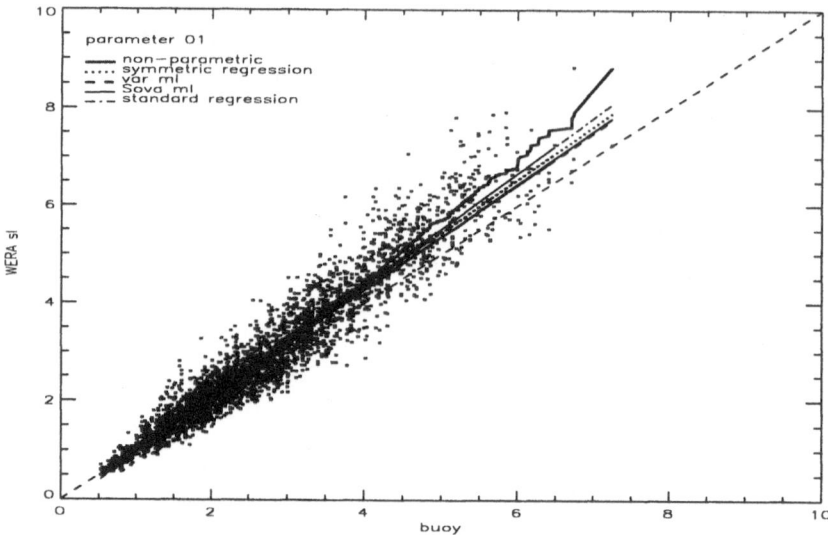

Fig. 6.4. As Fig. 6.3 for the Fedje measurements

data at high values of H_s. The difference in the exact value of H_s for which there appears to be a change in relationship might be explained by the difference in wave conditions at the two locations. Petten is in the southern North Sea which is shallow and fetch-limited, both of which features influence the shape of the spectrum. Fedje is in the North Atlantic in deep water exposed to Atlantic storms. The reason for the change in slope becomes apparent when one looks at the time series of H_s (Figure 6.5) and the way in which different frequencies contribute to it. In particular it can be seen that the higher frequency part of the spectrum (0.2-0.3Hz) is significantly overestimated in high seas and results in an overestimation of H_s above about 3m. This is demonstrating an inadequacy in the scattering theory that underpins the measurement.

The standard statistics for the H_s comparisons together with the slopes of the lines are presented in Table 6.1. The mean and mean square errors are very similar for the two experiments although a qualitative assessment of the scatter plots and the maximum likelihood analysis suggests that the Fedje HF measurements are in better agreement with the buoy. As has already been said these are not particularly helpful statistics. The differences in mean relative error are reflected in at least some of the slopes of the lines (a) – (d) but the standard regression slopes are similar in both cases. The qualitative differences seen by eye do therefore support the need to account for different error variances in both measurements.

Figure 6.6 shows time series comparisons of mean direction for the full spectrum, contribution from the spectral peak and from high frequency waves.

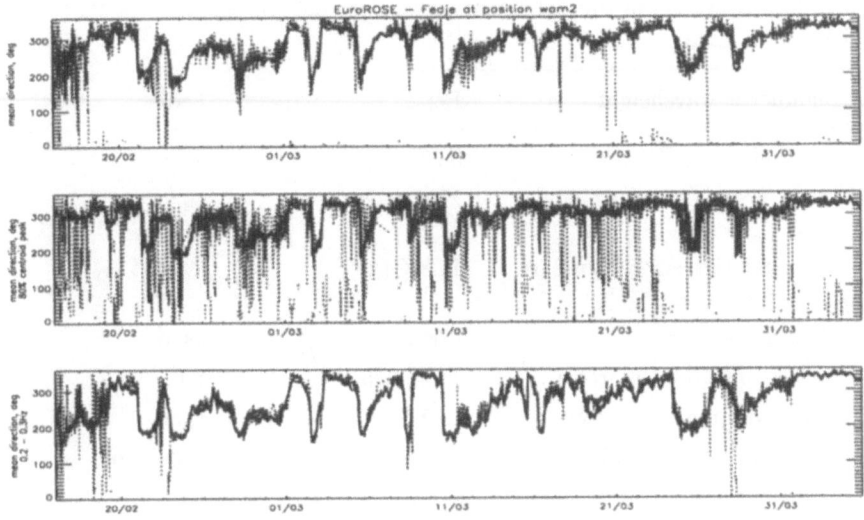

Fig. 6.5. Time series of significant waveheight (upper panel), contribution to H_s from spectral peak (middle panel) and from high frequencies (lower panel) at Fedje. Solid line is the buoy data and dotted line the radar data.

Statistics for these comparisons are in Table 6.2. It is clear that the mean difference is not particularly helpful since it is similar for the peak and high frequency bands but it is clear from the figure that the comparison is very different.

The other statistics provide a better measure of the comparisons. The main reason for the differences in peak direction shown here is presented in Figure 6.7 which shows a directional spectrum measured by the radar compared with the buoy. The peak in the buoy spectrum is at ∼0.1Hz and ∼140° and this feature can also be seen in the radar spectrum. However the radar spectrum also includes a component at ∼0.1Hz and 320° and the resulting mean direction in this frequency band is therefore significantly in error. The additional component is thought to be due to radar antenna sidelobe problems (Kingsley *et al.* 1997). The quantitative analysis of the directional spectra comparisons in Tables 6.1 and 6.2 and Wyatt *et al.* (1999) used fixed frequency bands. It would be more useful to partition spectra, that is decompose them into, for example, swell and wind-wave contributions or modes (Isaac & Wyatt 1997). The maximum likelihood methods could then be applied to the intercomparison of 'modes' particularly since this would reduce the impact of sidelobe contamination of the spectra. This is itself a statistical problem and we have not yet implemented an automatic procedure to carry this out.

The SCAWVEX results also demonstrate that there is variance in the data not explained by sampling variability (Krogstad *et al.* 1999) and therefore not captured in the variance estimate used in the generation of line (c). Caires

Table 6.1. Statistics of the significant waveheight intercomparisons at Petten and Fedje.

Significant waveheight	Petten	Fedje
Correlation coefficient	0.92	0.96
Mean error (standard deviation)	21cm (22.5%)	19cm (16%)
Mean relative error (standard deviation)	11.8% (24%)	6% (14.7%)
Mean square error	47cm	48cm
Regression slope/intercept	1.14/-0.05	1.14/-0.19
symmetric regression slope	1.13	1.09
Maximum likelihood slope/intercept with sampling variability variance estimates	1.05/0.02	1.09/-0.08
Maximum likelihood slope/intercept with variance estimates from data	1.13	1.08

Table 6.2. Statistics of the direction comparisons at Fedje.

Direction	Mean over all frequencies	Peak direction	Mean over 0.2-0.3Hz
T-linear correlation	0.81	0.38	0.85
Mean difference (standard deviation)	1.9° (27.6°)	6.7° (56.6°)	6.9° (25.9°)
Mean square error	27.7°	56.7°	26.8°

(2000) has developed a method to estimate variance from the data using a combined maximum-likelihood/pseudo-replication algorithm approach. This method can be (and has been) applied to more complex intercomparisons e.g. comparisons with wave models where Monte Carlo simulations are far too computationally expensive. This work seems to show that antenna sidelobes play a significant role in degrading the wave measurement capabilities of the HF radar away from the centre of the measurement region. Line (d) was generated using variances estimated with this technique. As has already been noted, at Fedje this gives a similar result to the other functional relationship line but this is not the case at Petten. In both cases the variances estimated are significantly larger then those associated with sampling variability both

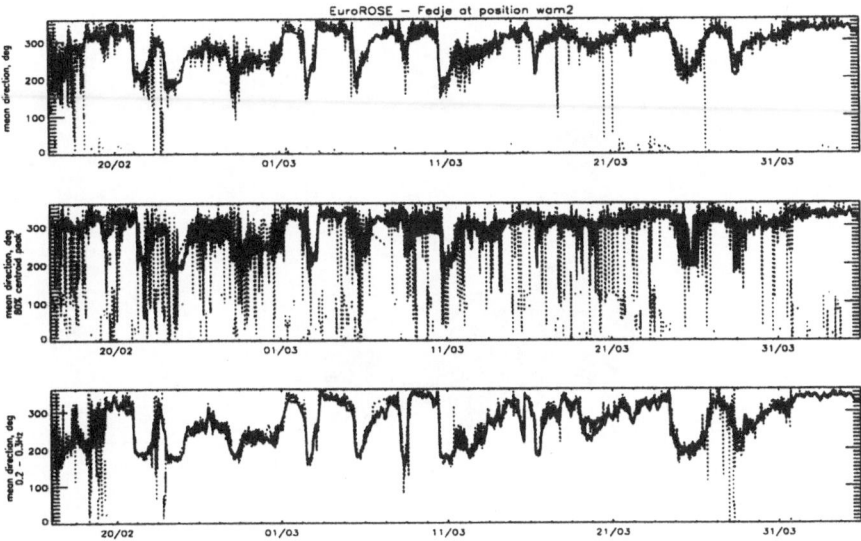

Fig. 6.6. Time series of mean direction (upper panel), spectral peak direction (middle panel) and mean high frequency direction (lower panel) at Fedje. Solid line is the buoy data and dotted line the radar data.

for the radar and, more surprisingly, the buoy. The differences between the two suggest that there are different origins for the variances which may again be associated with the different conditions at the two sites. In both cases the mean relative error estimate is similar to the maximum likelihood slope obtained using data-estimated variances.

6.6   Concluding Remarks

The purpose of this paper has been to show the sorts of problems that arise in validating new technologies for measurement of ocean surface waves against existing, well-established but not necessarily perfect, measurement systems. The statistical techniques that are used in HF radar and wave buoy comparisons have been described. I have tried to stress the need to combine different qualitative and quantitative techniques to identify weaknesses and strengths in different measurements. Individual statistics can provide a misleading interpretation of differences between two measurements.

As far as the HF radar and wave buoy intercomparisons are concerned, the analysis has revealed the main problems with the HF radar measurements. There are the contamination from antenna sidelobes leading to noisy peak direction estimates and the inadequacy in the scattering theory used to derive the equations that are inverted to provide the directional spectrum. Both these aspects are under study.

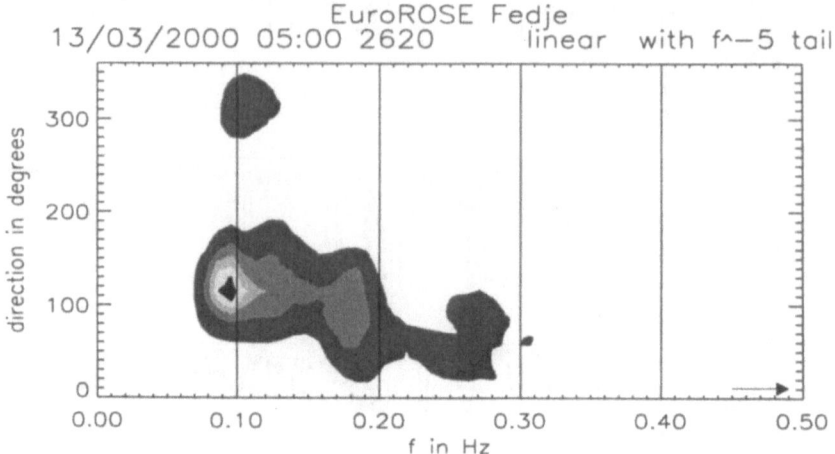

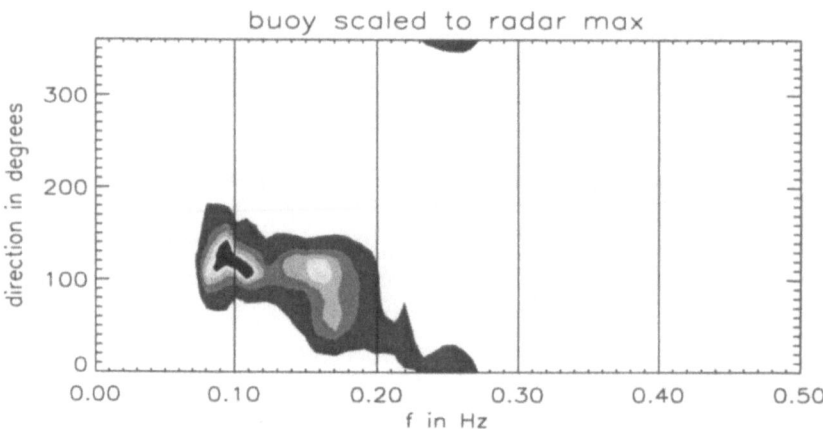

Fig. 6.7. Directional spectra measured by the radar (above) and wave buoy (below) during the EuroROSE Fedje experiment. Six grey scaled levels are shown with the darkest denoting the location of the peak. The arrow in the upper plot is the radar measured wind direction. The presentation is different from Fig. 6.2, both are commonly used. These plots show swell waves with a peak frequency of about 0.09Hz propagating from the north-west and a second system in roughly the same direction with a peak at about 0.16Hz. The energy in the wind direction is much lower than the swell energy.

Acknowledgements

Thanks are due to Markus Sova, Sofia Caires, Steve Thompson, Jim Green, Louise Ledgard (presently or previously of the University of Sheffield) and Harald Krogstad, Trondheim, Norway for various contributions and to Klaus-

Werner Gurgel and his group at the University of Hamburg who built the WERA radar and collected all the WERA data. Aspects of the work have been funded by the EU (SCAWVEX (MAS2CT940103), EuroROSE(MAS3CT 980168)), EPSRC and NERC.

References

Atanga, J. & Wyatt, L. R. (1997). Comparison of inversion algorithms for HF radar wave measurements. *IEEE Journal of Oceanic Engineering*, **22**, 593-603.

Bauer, E. & Heimbach, P. (1999). Annual validation of significant wave heights of ERS-1 synthetic aperture radar wave mode spectra using TOPEX/Poseidon and ERS-1 altimeter data. *Journal of Geophysical Research-C*, **104**, 13345-13357.

Borge, J. C. N., Reichert K. & Dittmer, J. (1999). Use of nautical radar as a wave monitoring instrument. *Coastal Engineering*, **37**, 331-342.

Caires, S. (2000). *Comparative study of HF radar measurements and wave model hindcasts in shallow waters*. PhD thesis, University of Sheffield.

Capon, J. (1969). High resolution frequency wavenumber spectrum analysis. *Proceedings of the IEEE*, **57**, 1408-1418.

Donelan, M. A., Drennan W. M. & Magnussen, A.K. (1996). Nonstationary analysis of the directional properties of propagating waves. *Journal of Physical Oceanography*, **26**, 1901-1914.

Günther, H., Gurgel, K.-W., Evensen, G., Guddal, J., Nieto Borge, J.-C. & Wyatt, L. R. (2000). European radar ocean sensing. *Proceedings of EurOCEAN 2000*, Hamburg, 2000, 443-448.

Hashimoto, N. (1997). Analysis of the directional wave spectrum from wave data. *Advances in Ocean and Coastal Engineering*, **3**, 103-143.

Hasselman, K. & Hasselman, S. (1991). On the non-linear mapping of an ocean wave spectrum into a Synthetic Aperture Radar image spectrum and its inversion. *Journal of Geophysical Research-C*, **96**, C6, 10713-10729.

Junger, S. (1998). *The Perfect Storm*. Fourth Estate, London.

Isaac, F. E. & Wyatt, L. R. (1997). Segmentation of HF radar measured directional wave spectra using the Voronoi Diagram. *Journal of Atmospheric and Oceanic Technology*, **14**, 950-959.

Kingsley, S. P., Blake, T. M., Fisher, A. J., Ledgard, L. J. & Wyatt, L. R. (1997). Dual HF radar measurements of sea waves from straight coastlines. *Proceedings of the 7th International Conference on HF radio systems and techniques*, Nottingham, July 1997 (IEE, London UK).

Krogstad, H. E., Wolf, J., Thompson, S. P. & Wyatt, L. R. (1999). Methods for the intercomparison of wave measurements. *Coastal Engineering*, **37**, 235-258.

Krogstad, H. E. & Barstow, S. F. (1999). Satellite wave measurements for coastal engineering applications. *Coastal Engineering*, **37**, 283-308.

Lygre, A. & Krogstad, H. E. (1986). Maximum entropy estimation of the directional distribution in ocean wave spectra. *Journal of Physical Oceanography*, **16**, 2052-2060.

Sova, M. G. (1995). *The sampling variability and the validation of high frequency wave measurements of the sea surface*. PhD thesis, University of Sheffield.

Tucker, M. J. (1991). *Waves in Ocean Engineering: Measurement, Analysis, Interpretation*. Ellis Horwood series in Marine Science.

Wyatt, L. R. (2000). Limits to the inversion of HF radar backscatter for ocean wave measurement. *Journal of Atmospheric and Oceanic Technology*, **17**, 1651-1666.

Wyatt, L. R. & Holden, G. J. (1992). Developments in ocean wave measurement by HF radar. *IEE Proceedings-F*, **139**, 170-174.

Wyatt, L. R. & Holden, G. J. (1994). Limits in direction and frequency resolution for HF radar ocean wave directional spectra measurement. *The Global Atmosphere and Ocean System*, **2**, 265-290.

Wyatt, L. R. & Ledgard, L. J. (1996). OSCR wave measurement – some preliminary results. *IEEE Journal of Oceanic Engineering*, **21**, 64-76.

Wyatt, L. R. (1990). A relaxation method for integral inversion applied to HF radar measurement of the ocean wave directional spectrum. *International Journal of Remote Sensing*, **11**, 1481-1494.

Wyatt, L. R. (1991). HF radar measurements of the ocean wave directional spectrum. *IEEE Journal of Oceanic Engineering*, **16**, 163-169.

Wyatt, L. R. (1995). High order nonlinearities in HF radar backscatter from the ocean surface. *IEE proceedings–Radar, Sonar and Navigation*, **142**, 293-300.

Wyatt, L. R., Venn, J., Moorhead, M. D., Burrows, G. D., Ponsford, A. M.& van Heteren, J. (1986). HF radar measurements of ocean wave parameters during NURWEC. *IEEE Journal of Oceanic Engineering*, **OE-11**, 219-234.

Wyatt, L. R., Gurgel, K-W., Peters, H. C., Prandle, D., Krogstad, H. E., Haug, O., Gerritsen, H., Wensink, G. J. (1998). The SCAWVEX Project: ocean wave measurement and analysis. *Proceedings of WAVES 97*, eds. Edge, B. L.& Hemsley, J. M., vol **2**, 1457-1467.

Wyatt, L. R., Ledgard, L. J. & Anderson, C. W. (1997). Maximum likelihood estimation of the directional distribution of 0.53Hz ocean waves. *Journal of Atmospheric and Oceanic Technology*, **14**, 591-603.

Wyatt, L. R., Thompson, S. P. & Burton, R. R. (1999). Evaluation of HF radar wave measurement. *Coastal Engineering*, **37**, 259-282.

7 Thermal Energy Emission and Propagation from Accidents

A. Pelliccioni, F. Altavilla, and S. Berardi

ISPESL (National Institute for Occupational Prevention and Safety),
Department of Production Plants and Interaction with the Environment,
Via Fontana Candida,1 – 00040 Monteporzio Catone, Roma, Italy

Abstract. Many different kinds of accident may occur inside an industrial plant. Such events often produce, directly or indirectly, thermal radiation. The intensity of this radiation is frequently so high (for example, in fires and boiling liquid expanding vapour explosions – BLEVEs) that it causes significant damage to industrial installations or to people.
For industrial application, the calculation of the intensity of thermal radiation incident on a surface placed at a certain distance from a source is traditionally carried out by considering only the contribution of the radiation directly hitting the surface. Thus, in accident modelling, contributions due to radiation reflected from the ground, or absorbed by the ground and subsequently re-emitted, are neglected. The present work describes a model able to take into account these three contributions to the thermal radiation resulting from an accident.
The model calculates the energy re-emitted by the ground using a diffusion equation in one or two spatial dimensions. The space-time distribution of temperature, assuming the ground thickness is infinite both in depth and in extension, is estimated. The model is applied under different scenarios involving thermal radiation emission. The results show that the two contributions of thermal energy give additional effects that cannot be neglected in assessing risk exposure.

7.1 Introduction

This study analyses the phenomenon of the absorption and subsequent re-emission of thermal energy incident on a surface following an accident. In particular, the case of a Boiling Liquid Expanding Vapour Explosion (BLEVE) is studied. A BLEVE occurs for example when a tank containing liquefied petroleum gas (LPG) is exposed to fire, fails catastrophically and an immediate fireball occurs (Papazoglou & Aneziris 1999). The thermal radiation produced from the fireball is always very intense and, although the phenomenon has a short life, it may cause very serious effects to human health, property and the environment up to hundreds of metres from the accident site.

For risk management and emergency planning purposes, steady-state and non-steady heat radiation limit values are defined in order to compute the relevant distances at which preventive and protective actions should be provided. For example, Table 7.1 presents the threshold values adopted for this purpose by the Dipartimento della Protezione Civile (Department of Civil Protection) in Italy (DPC 1994). Two kinds of phenomena are considered in Table 7.1: steady-state phenomena linked to the thermal energy flux and measured in Wm^{-2}, and non-steady phenomena linked to the thermal dose and measured in Jm^{-2}. To change from one condition to another the phenomenon duration must be taken into account. For instance, in the case of a fireball following a BLEVE, having a lifetime of 10 seconds, a dose of 125 Jm^{-2} is equivalent to a flux of 12.5 kW m^{-2}. (This conversion is, of course, over-simple since it assumes constant conditions over the lifetime.)

Table 7.1. Receptor damage threshold due to radiation and explosion

Physical Phenomenon	High lethality	Beginning lethality	Non-reversible injury	Reversible injury	Structural damage, domino effects
FIRE (steady-state thermal radiation)	12.5 kWm^{-2}	7 kWm^{-2}	5 kWm^{-2}	3 kWm^{-2}	12.5 kWm^{-2}
BLEVE/Fireball (non-steady thermal radiation)	Fireball radius	350 kJm^{-2}	200 kJm^{-2}	125 kJm^{-2}	From 100 to 800 m

However, such safe distances are valid merely for the duration of the phenomenon and they are closely linked to the consequences of the accident. Potentially important thermal contributions are due either to reflected radiation or radiation absorbed by the ground and subsequently re-emitted. The aim of the present paper is to reach a qualitative assessment of the increase of available energy in the system due to these ground effects. In this first study we assume throughout that the ground is a uniform smooth surface. For more complicated surfaces, involving for example variable texture, the calculations in this paper can be integrated when appropriately weighted to provide the net effect on a target volume.

In thermal effect assessment, usually only the source emissive power directly hitting a target placed at a certain distance from the event is considered, and this direct flux is accounted as exhausting the effects of the phenomenon, neglecting either the contribution of energy reflected from the ground or the part of energy absorbed by ground and subsequently re-emitted. Figure 7.1 shows the scheme used here to estimate the contribution from these further effects. The fireball is the main energy flux source. Three kinds of energy fluxes fall on an infinitesimal target volume parallel to the ground. One is the direct energy coming from the explosion, the other two are new contributions derived from the ground, and consist of reflected and re-emitted energy.

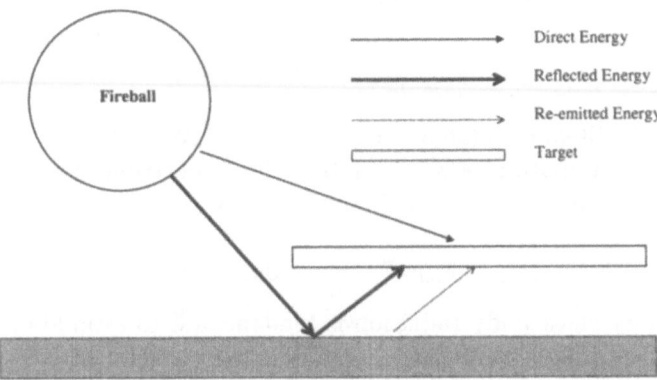

Fig. 7.1. Main energy contributions incident on a target volume

The study demonstrates that a high intensity thermal energy source hitting a surface, even if for a very short time (a fireball does not exceed 20 seconds, generally) causes a modification of surface thermal conditions that can be felt some time after the event has exhausted itself. Computations show that, after such an event, the system temperature rapidly increases and subsequently slowly cools. The paper describes the global radiative model employed to obtain these results and goes on to report the results of a study of the effect of ground surfaces of different material. Further applications related to dose and to safe distances are also described.

7.2 Heat Transfer Theory

To carry out global radiative modelling, we first consider general heat transfer theory. Subsequently we report the criteria, assumptions and theoretical formulations adopted to assess the thermal response of the ground following an unsteady thermal accident event. To assess the heating effects on the ground following an accident it is necessary to link different topics related to radiation emitted by a black body and heat transfer.

7.2.1 Black Body Emissive Power Calculation

The power spectrum of radiation emitted by a surface at temperature T (K) is given by Planck's Law (see, for example, Hunt 1980):

$$W(\lambda, T) = 2\pi hc^2 \lambda^{-5} / \left\{ \exp\left(\frac{ch}{\lambda kT}\right) - 1 \right\}, \qquad (7.1)$$

where λ denotes the wavelength, c is the velocity of light, h is Planck's constant and k is Boltzman's constant.

Equation (7.1), integrated over all wavelengths, leads to the Stefan-Boltzman law:

$$W = \int_0^\infty W(\lambda)d\lambda = \sigma T^4,$$

where σ is the Stefan-Boltzman factor ($5.56.10^{-8}$ W m^{-2} K^{-4}). This expression gives the radiating power over the whole spectrum. The wave length, λ_{max}, corresponding to the maximum emission power is calculated by the Wien law:

$$\lambda_{max} T \quad = \quad 2890 mK \tag{7.2}$$

The theory of black body radiation is fundamental to estimation of the energy fluxes entering and leaving the ground. We consider three input energy sources: the sun; the fireball developing after the BLEVE; and radiation from the atmosphere. We take account of the change in absorption properties of materials in relation to the spectrum of the incident energy. Furthermore, the model assumes that the ground cools through emitting an amount of energy in accordance with its absorption factor and the Stefan-Boltzman law.

7.2.2   Energy Fluxes from a Fireball

The calculation of the energy fluxes coming from a fireball is carried out by applying the surface-source model (TNO 1992). The heat load received by the ground, per unit surface area, is given by:

$$P(r) = \tau_a F_v \varepsilon E_t \tag{7.3}$$

in which τ_a is atmospheric transmissivity, F_v is the view factor between the unit surface and emissive source, ε is the radiation source emittance and E_t is the total emissive power of a black radiator at the fireball temperature. We used the formulation adopted by TNO (1992) that is incorporated in the EFFECTS2 code (TNO 1996).
The view factor is given by:

$$F_\nu = \left(\frac{r}{x}\right)^2 \cos \vartheta,$$

where r is the fireball radius, x is the distance measured on the ground between the receptor and the centre of the fireball and ϑ is the angle between the normal to the receiving surface and the line from the receptor to the centre of the fireball. The fireball radius, r, in metres and the fireball duration, t, in seconds are estimated by the following empirical formulae (TNO 1983):

$$r = 3.24 m^{0.325}$$

$$t = 0.852 m^{0.26} \tag{7.4}$$

where m is the mass in kilograms of the fireball.

7.2.3   Albedo

The flux of radiating energy incident on the ground surface is partly absorbed and partly transmitted and/or reflected from the ground. For every wavelength, the following formula, following from conservation of energy, is valid:

$$\alpha(\lambda) + \beta(\lambda) + \tau(\lambda) = 1, \tag{7.5}$$

where α, β and τ are respectively the absorption, reflection and transmission factors. These factors are functions of the chemical and physical nature of the irradiated body, of its surface roughness and of the incident radiation wavelength. In our model structure, we have assumed an infinite ground thickness. Thus the transmission factor τ is equal to zero, and equation (7.5) becomes:

$$\alpha(\lambda) + \beta(\lambda) = 1.$$

The reflectance factor $\beta(\lambda)$ for each surface can be derived from literature data or can be estimated experimentally using spectro-radiometers (Colwell 1983). In Figure 7.2 typical factors of some materials commonly used in industrial paving are shown. For accidental scenarios, it is not so important to

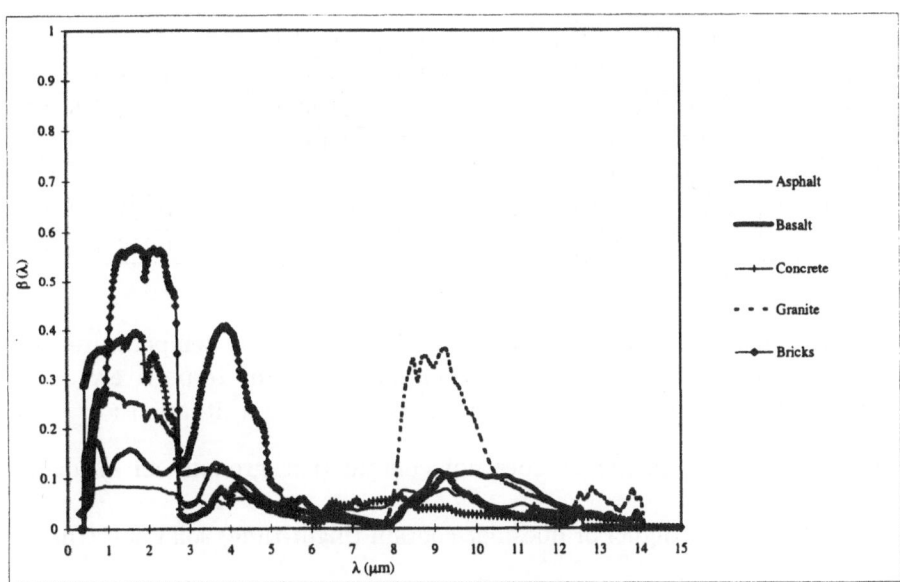

Fig. 7.2. Surface reflection factor

know exactly the reaction of the surface of the system at each wavelength, but the integral response, that is, the albedo. The albedo is defined as the ratio between the electromagnetic energy, reflected or scattered by the surface, and the total incident energy (Colwell 1983). In our model, we have derived the following expression for albedo a:

$$a = \frac{\int_0^\infty \sum_{k=1}^{N} W(\lambda; T_k)\beta(\lambda)d\lambda}{\int_0^\infty \sum_{k=1}^{N} W(\lambda; T_k)d\lambda}. \tag{7.6}$$

In (7.6), $W(\lambda; T_k)$ is the emission spectrum of incident energy sources at temperature T_k, from Planck's law (7.1). Note that, in our application, the integrals are evaluated over wavelengths between $0.2\mu m$ (λ_{min}) and $15\mu m$ (λ_{max}) consistent with the material surfaces under consideration (Figure 7.2). For the emission spectra at different temperatures of the three sources ($T = 6000K$ for the sun, $T = 1460K$ for the fireball and $T = 300K$ for the atmosphere), the albedo values given by (7.6) for some typical surfaces are listed in Table 7.2.

Table 7.2. Surface albedo values

SURFACE	Background (day)		Accident (day)		Background (night)		Accident (night)	
	a (partial)	a (total)	a (partial)	a (total)*	a (partial)	a (total)	a (partial)	a (total)
Asphalt	0.077	0.074	0.077	0.074	0.038	0.021	0.065	0.064
Concrete	0.294	0.282	0.293	0.281	0.030	0.016	0.184	0.182
Water	0.001	0.001	0.001	0.001	0.016	0.008	0.015	0.015
Bricks	0.240	0.230	0.241	0.231	0.037	0.018	0.361	0.356
Basalt	0.152	0.130	0.151	0.130	0.053	0.028	0.097	0.095
Granite	0.202	0.194	0.202	0.194	0.105	0.056	0.152	0.150

The maximum emissive power values at the three source temperatures are, from the Wien law (7.2), $0.48\ \mu m$, $1.98\ \mu m$ and $9.73\ \mu m$ respectively. These data have been calculated to assess two radiation situations:

- it contains (accident) or does not contain (background) an incidental event;
- it contains (daylight) or does not contain (night-time) solar spectrum.

Moreover, each radiation situation has been evaluated according to the integration range of equation (7.6):

- total: integration wavelength interval between 0 and $+\infty$
- partial: integration wavelength interval between λ_{min} and λ_{max}.

Table 7.2 shows that the albedo values of the surfaces examined are affected more by the solar spectrum than by other factors.

7.2.4   Heat Transfer Formulations

Some approximations have been made to simplify the system dynamics and to harmonize the mathematical formalism with reality.

The first approximation is to assume spherical symmetry of the fireball. We made another approximation, linked to the physical characteristics of the system (specific heat c_p, density ρ, heat radiation coefficient λ), which consists in assuming that these quantities take constant temperature values ($T = 300K$). This approximation is consistent with the consideration that the heat load supplied to the system does not cause large modification in the physical behaviour of materials.

Finally, it has been assumed that the irradiated surface has spatially homogeneous chemical and physical properties. Simulations have been performed on a two-dimensional space, having horizontal axis (x) and vertical axis (z). The Fourier (or diffusion) heat equation (see, for example, Incropera & de Witt 1996) has been solved to calculate the temperature following an event. The following formulation has been adopted in the model:

$$\frac{\lambda_h}{c_p\rho}\left(\frac{\partial^2 T}{\partial x^2} + \frac{\partial^2 T}{\partial z^2}\right) = \frac{\partial T}{\partial t}. \tag{7.7}$$

Considering the heat radiation transfer as a heat flux supplied to the body volume and adopting the Neumann boundary condition (Fletcher 1991), the heat fluxes Φ_s incident on the surface produce temperature gradients calculated by the following equation:

$$\frac{\partial T}{\partial z} = -\frac{\Phi_s}{\lambda_h},$$

where λ_h is the heat radiation coefficient. A finite difference method (Fletcher 1991) is used to solve the diffusion equation at the boundary.

7.2.5   Ground Cooling and Restoring Time

The model enables us to calculate the ground cooling and the time that the system needs to regain its pre-incident temperature. This temperature depends in a complex way on the physical characteristics of the ground and on the boundary conditions linked to fluxes. As a result it is not possible to estimate the time necessary to regain the ground background temperature from the simulation without a very long calculation time.

It has been assumed that the ground cooling occurs according to a polynomial law:

$$T(x_0; t) = \gamma(x_0)t^{\delta(x_0)}, \tag{7.8}$$

where $\gamma(x_0)$ and $\delta(x_0)$ are empirical parameters to be calculated for a point x_0 on the ground following a temperature trend decreasing with time according to the simulation. The formulation (7.8) is derived empirically from tests on the radiative cooling of different materials subject to incident thermal energy. The time t_{eq} required to restore the temperature T_{eq} is given by:

$$t_{eq} = \left(\frac{T_{eq}}{\gamma(x_0)}\right)^{1/\delta(x_0)} - t_{inc} \qquad (7.9)$$

where t_{inc} is the duration of the event. Equation (7.9) can also be used to estimate the time to reach a specified temperature, perhaps of interest for safety and health reasons, and to establish safe admittance to an area affected by an accident.

7.3   Global Radiative Model Architecture and Application

The model contains three main calculation modules.
The first module calculates the albedo of the surface receiving the heat fluxes. As input, the model requires the emission spectrum of the surface and the source temperature. The second module takes as input the albedo value, the physical properties of the surface (c_p, ρ and λ_h) and the thermal flux incident on the surface; this module provides the portion of energy reflected by the ground, the energy absorbed and subsequently re-emitted and the space-time distribution of temperature. The third module evaluates the time necessary for the system to re-establish the thermal conditions before the accident. In order to do that, the temperature T_{eq}, the parameters γ and δ and the time at which the event began are needed as input in equation (7.9).
As an example we select a BLEVE/fireball scenario for a tank containing 25,000 kg of liquefied propane gas with a subsequent atmospheric release of about 10,000 kg. The intensity of thermal radiation caused by the explosion is given by equation (7.3) as a function of distance from the surface to the source of solid-flame (Figure 7.3). The heat flux at ground level is about 49 kW m^{-2} at the origin ($x = 0$m) and decreases rapidly with x. The explosion duration, found from equation (7.4), is 9.34 seconds.
The present simulation uses the output of the EFFECTS2 code (TNO 1996) and the diffusion equation (7.7). The conceptual model of the receiving ground is two-dimensional; its maximum horizontal length is 1513.8 m and its depth is 0.5m. This rectangle is divided into a grid with meshes $x = 34.5$m and $z = 0.025$m. The initial temperature at each knot of the grid was taken equal to 300K. The ground has been supposed to be of different kinds of paving materials or substances (asphalt, water, concrete, bricks, basalt, granite). The model input may specify materials whose physical characteristics (Nuovo Columbo 1985) are listed in Table 7.3, with the corresponding albedo values from (the fifth, asterisked, column of) Table 7.2.
The simulation results show that the maximum ground temperature is reached at the end of the duration of the fireball (about 10 seconds) at every point

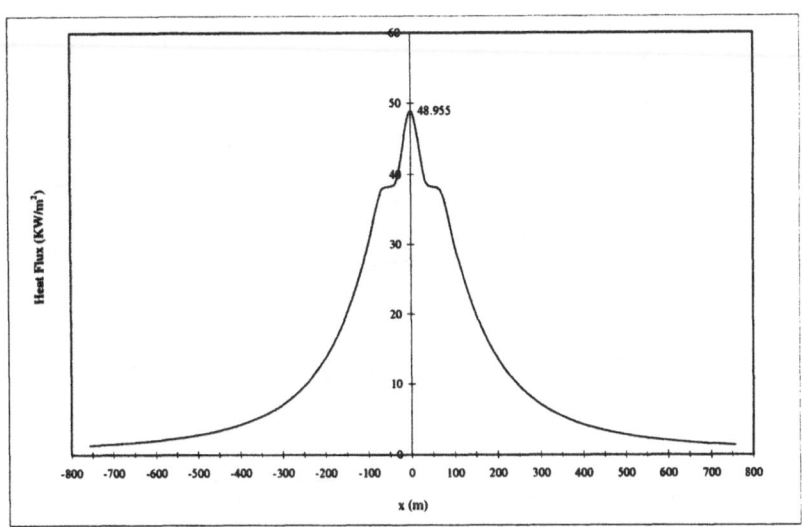

Fig. 7.3. Intensity of thermal radiation as a function of distance from a BLEVE

Table 7.3. Surface physical characteristics

SURFACE	$\lambda(W/m\ ^\circ K)$	ρ (kg/mc)	c_p (J/kg $^\circ K$)
Asphalt	0.64	1300	933
Concrete	1.16	2500	1134
Water	0.60	1000	4184
Bricks	0.59	1600	860
Basalt	1.50	1900	879
Granite	3.20	2500	879

of the surface and for every material. After the explosion ($t > 9.34s$), the thermal flux away from the surface results in the surface returning to background temperature very gradually. The thermal trend at two representative surface locations, the point at the origin of the fireball ($x = 0.0$m) and a point just outside the fireball radius ($x = 65.0$m), are shown in Figure 7.4 and Figure 7.5.

The figures show features which are well-known, such as the systematic decrease of the temperature with distance from the point-source, the sudden elevation of temperature due to the explosion and its subsequent decrease. However, this temperature decrease is not the same for all materials. The maximum temperature ($T_{\max}$), incident ground power (P_{tot}), absorbed ground

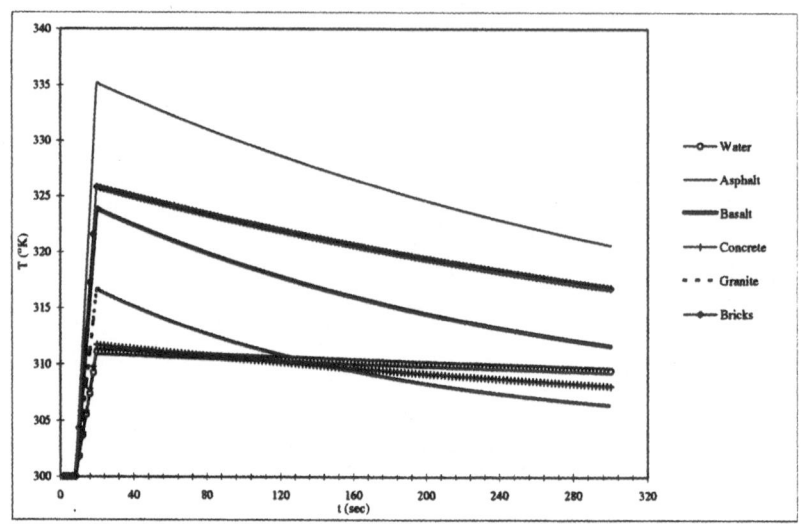

Fig. 7.4. Temperature trend at $x = 0.0$m

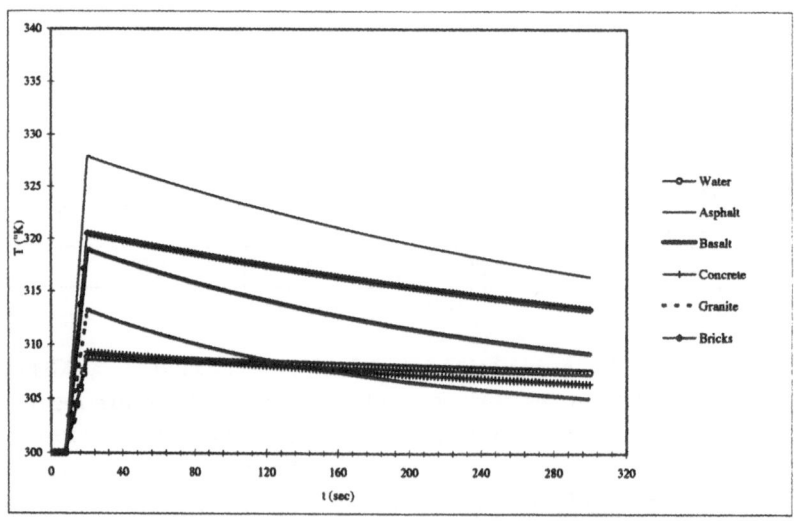

Fig. 7.5. Temperature trend at $x = 65.0$m

power (P_{abs}), reflected ground power (P_{ref}) and re-emitted ground power at the end of the fireball (P_{re-em}) are listed in Table 7.4 for the two distances on the ground and for each material. The re-emitted ground power (P_{re-em}) is calculated from:

$$P_{re-em} = (1 - a)\sigma T_{max}^4.$$

Table 7.4. Temperature and power values at $x = 0.0$m and $x = 65.0$m

$t = 10\,\mathrm{s}\;;\; x = 0.0\,\mathrm{m}$

SURFACE	T_{max} ($^\circ K$)	P_{tot} (W/m^2)	P_{abs} (W/m^2)	P_{ref} (W/m^2)	P_{re-em} (W/m^2)
Asphalt	335	48955	45332	3623	714
Concrete	311	48955	35150	13805	530
Water	311	48955	48906	49	530
Bricks	326	48955	37695	11260	640
Basalt	324	48955	42591	6364	625
Granite	317	48955	39458	9497	573

$t = 10\,\mathrm{s}\;;\; x = 65.0\,\mathrm{m}$

SURFACE	T_{max} ($^\circ K$)	P_{tot} (W/m^2)	P_{abs} (W/m^2)	P_{ref} (W/m^2)	P_{re-em} (W/m^2)
Asphalt	328	38920	36040	2880	656
Concrete	309	38920	27945	10975	517
Water	309	38920	38881	39	517
Bricks	320	38920	29968	8952	595
Basalt	319	38920	33860	5060	587
Granite	313	38920	31370	7550	544

Generally, the thermal behaviour of bodies is not uniquely determined by their heat propagation properties (linked to their physical and chemical characteristics), but depends also on the energy absorbed by the body, which is a function of albedo.

We have verified that the value of the albedo is not critical when the heat transfer inside a body is to be calculated, but it is essential in the calculation of energy incident on a target volume (Figure 7.1). In fact, an interesting finding from the simulation is the minimal difference in thermal reaction of materials having similar physical characteristics. The percentage difference between asphalt and brick physical characteristics (c_p, ρ and λ) is about 12% and the albedo values are 0.074 and 0.231 respectively (which are very different), but the maximum temperature difference between the two materials is only about 9K. On the other hand, water and asphalt, which have similar albedo values and different physical characteristics, have differing thermal reactions: the maximum temperature reached by water is 311K, while that for asphalt is 335K.

Apart from the above considerations, it is possible to estimate the effective thermal reaction of different materials, subject to the same thermal excitation, through knowledge of both physical characteristics and optical properties of the material surface. Furthermore, it is interesting to note that the power re-emitted by the ground (Table 7.4) can reach significant exposure levels: under the BLEVE ($x = 0$), the minimum value is 530 Wm^{-2} and the maximum value is 714 Wm^{-2} (which correspond to increases of 15.5% and of 55.5% compared to the background of 459 Wm^{-2}). It is to be emphasized that commonly used theoretical approaches neglect this contribution.

Finally, the return period necessary to reach the pre-explosion temperature is computed by equation (7.9). During the cooling phase, the ground will re-emit absorbed energy until it reaches its pre-explosion equilibrium temperature. This return period is computed by equation (7.5), adopting the value $T_{eq} = 300K$. The results are shown in Table 7.5. It should be noted that the return periods are very different for different materials. Asphalt, for example, is the warmest material at the end of the fireball and it cools rather quickly; in contrast the concrete return period is three times that for asphalt.

Table 7.5. Return period to pre-explosion background temperature $T_{eq} = 300K$

SURFACE	$x = 0.0$ m	$x = 65.0$ m
Asphalt	2383	2344
Concrete	5838	5231
Water	181838	180739
Bricks	3768	3590
Basalt	1720	1590
Granite	1398	1322

7.4   Comments on Model Results

7.4.1   Calculation of IP Index

Whenever we need to evaluate hazards connected to an incident, it is necessary to relate the phenomenon intensity to the expected degree of injury or damage. In the case of a thermal energy release, the effects are connected to the emitted energy intensity (Figure 7.4). In particular, to analyse the effects on human beings we must know the value of the emission intensity at a certain place and the duration of exposure to the corresponding radiation. A measure of the effect of a particular exposure is the dose value (TNO 1989), which is given by

$$d(r) = [P(r)]^{4/3}t,$$

where $P(r)$ is the thermal flux at a distance r, t is the exposure duration and $d(r)$ is the dose measured at a distance r. The thermal dose is measured in TDU (Thermal Dose Unit) - whose dimensions are $(Wm^{-2})^{4/3}$ s – and is a measure of intensity of the factor which harms the vulnerable resource.

In order to assess the consequences following a given dose absorption, we must first point out that the damage to individuals caused by a thermal accident is not always the same. In fact, generally, one subject reacts to the phenomenon in a different way from another. This aspect motivates a probabilistic rather than a deterministic approach to damage assessment. A quantitative measure of the effect on human health is the probability of loss of life, as specified by the Probit (Probability Unit) function, defined by the relationship (Lees 1986)

$$Y(r) = k_1 + k_2 \ln d(r),$$

where $Y(r)$ is the percentage of vulnerable resources which sustain injury or damage, and k_1 and k_2 are parameters distinctive of the particular accident. If it is assumed that within a certain population the Probit is a random variable with a mean 5 and a standard deviation 1 (Lees 1986), the loss of life probability $IP(r)$ following the absorption of dose $d(r)$ is given by the relationship

$$IP(r) = \frac{1}{(2\pi)^{1/2}} \int_{-\infty}^{Y(r)-5} \exp(-u^2/2)du.$$

This probability for the selected accident may be calculated by the traditional approach from the thermal power incident at the ground derived from the EFFECTS2 code (TNO 1996), on the basis of the corresponding determination of the dose. However, by taking into account exclusively the unsteady effects linked to the fireball duration, we can evaluate the further contributions of energy to the system due to the presence of the ground. In Figure 7.6 the thermal dose coming from the classical model by means of the EFFECTS2 code is shown. We can observe that the maximum value of thermal dose is about 1650 TDU, estimated beneath the fireball, decreasing considerably with distance. For each surface material considered, application of the model always shows a clear rise in percentage of the dose $d(r)$ with distance (Figure 7.7). In particular, referring to ground surfaces made from bricks and concrete, we observe a considerable increase of the dose, up to 40.3% at the point where the accident has occurred, increasing to over 75% at a distance of about 760 m from that point. It is seen that larger percentage increases in dose occur at greater distances from the explosion. However, these percentage increases are relative to a very low radiation level, and so correspond to insignificant real effects,

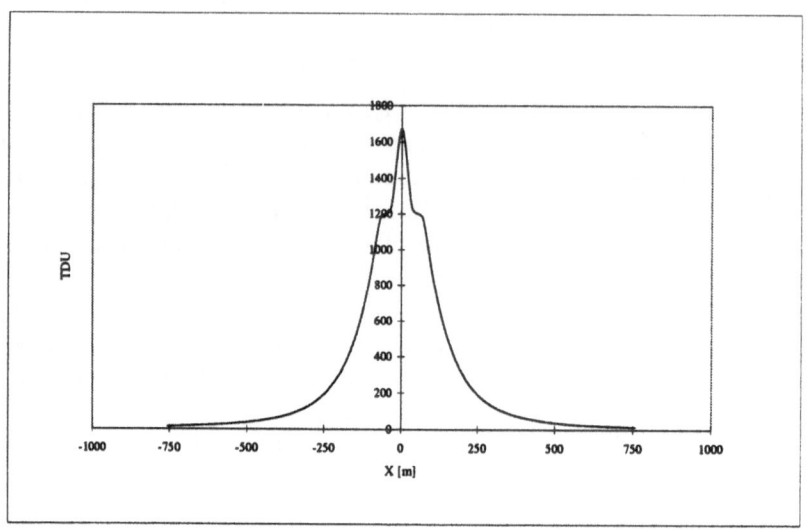

Fig. 7.6. Calculation of the dose coming from the classical model

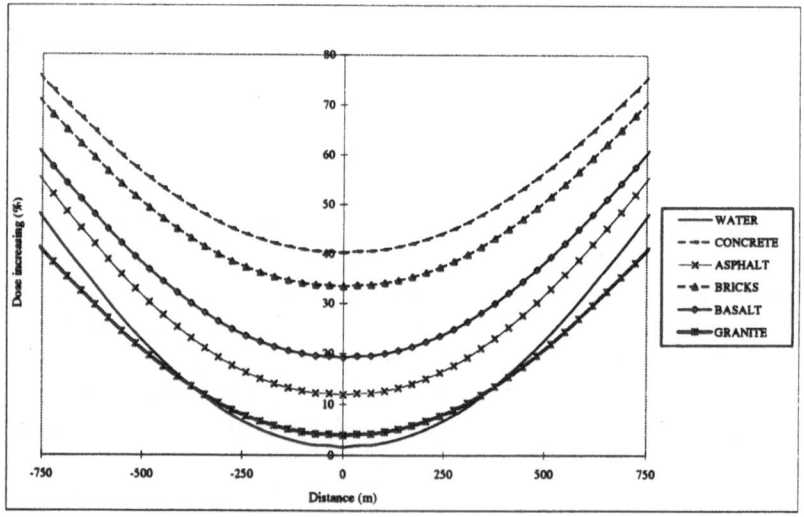

Fig. 7.7. Percentage increase in dose due to paving

Figure 7.8 shows the probability of loss of life during the duration of the explosion, as found by the classical approach. The loss of life expected is about 18.4% of the population beneath the centre of the fireball and its value decreases steeply to a value of 0.5% at the border of the fireball. Taking into account the dose increase on an individual due to the presence of the ground, changes in IP for different materials have been calculated (Table 7.6). In all cases selected, an increase in IP compared to the classical approach is seen. The increase is especially noteworthy in the case of concrete and brick paving: the expected percentage increase of the probability IP of loss of life below the explosion centre ($x = 0$) is 137% for bricks and 167% for concrete paving.

Table 7.6. Increase in probability IP calculated by the model with a ground surface compared to the classical model

	Distance x	0.0	34.4	68.8	103.2	137.6
	IP [%] Classical	18.4	4.7	3.7	0.5	0.0
Water	IP [%] Water	19.5	5.2	4.1	0.6	0.0
	Percentage Increase	5.9	10.6	11.5	11.1	0.0
Concrete	IP [%] Concrete	49.1	20.8	17.7	3.7	0.0
	Percentage Increase	166.9	343.7	384.0	641.8	0.0
Asphalt	IP [%] Asphalt	27.1	8.3	6.7	0.9	0.0
	Percentage Increase	47.5	77.4	84.0	85.0	0.0
Bricks	IP [%] Bricks	43.5	17.3	14.6	2.7	0.0
	Percentage Increase	136.7	269.3	299.1	453.3	0.0
Basalt	IP [%] Basalt	32.5	11.1	9.1	1.4	0.0
	Percentage Increase	76.6	136.1	148.9	175.9	0.0
Granite	IP [%] Granite	21.1	5.8	4.6	0.6	0.0
	Percentage Increase	14.6	23.6	25.4	22.4	0.0

7.4.2   Calculation of Safe Distance

To estimate safe distances, the Dipartimento della Protezione Civile (Department of Civil Protection) in Italy considers five different threshold levels

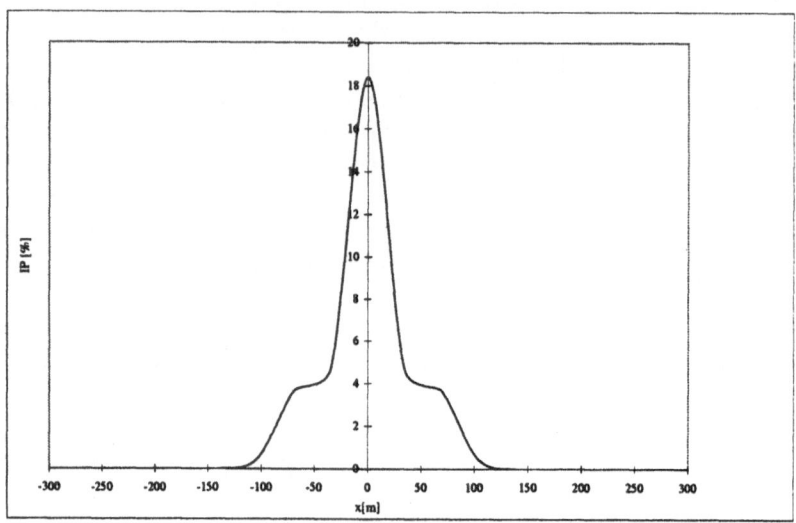

Fig. 7.8. Probability IP of loss of life (percentage) from the classical model

(Table 7.1). If we consider exclusively the unsteady effects limits, we can evaluate the impact distance with reference to three levels of injury or damage (reversible and non-reversible injury and beginning lethality), starting from the distribution on the ground of energy fluxes. The results, obtained from classical modelling, provide the safe distance values shown in Table 7.7. We note that the distances vary considerably according to the energy flux levels, rising from 69.2m (beginning lethality limits) to 203.2 m (reversible injury limits). When the presence of the ground surface is taken into account the safety distance for these three limits display a differentiated increase according to material (Table 7.7). In this table too the percentage increases compared to estimates from the classical approach are provided.

We note that the classical method underestimates distances by a percentage which increases with increasing damage, reaching the maximum value (50.4%) in the case of concrete paving and lethality limits. Of the materials considered, concrete is the material for which safe distances are most seriously underestimated (19.2%, 23.7% and 50.4% for the three levels respectively), followed by bricks and basalt.

7.5 Conclusions

Simulations have shown that a total incident power percentage up to 99.9 % is absorbed by the different materials, which re-emit this absorbed power

Table 7.7. Safe distances in metres due to unsteady effects

	Reversible injury		Non-reversible injury		Beginning lethality	
	Dist.	% Inc.	Dist.	% Inc.	Dist.	% Inc.
Classical approach:	203.2	–	142.0	–	69.2	–
Present model:						
Concrete	242.2	19.2	175.7	23.7	104.1	50.4
Bricks	237.1	16.7	170.6	20.1	99.2	43.4
Basalt	227.7	12.1	160.8	13.2	88.6	28.0
Asphalt	219.0	7.8	154.8	9.0	81.9	18.4
Granite	209.9	3.3	147.3	3.7	73.9	6.8
Water	207.7	2.2	149.9	5.6	71.8	3.8

gradually, with a percentage up to 28.1% being reflected instantaneously. Therefore, generally, these reflected and re-emitted energies can induce additional damage to the receptor receiving the direct emissive power. This aspect is of special interest when damage resulting from unsteady accidental scenarios (BLEVE) is studied. In fact, this accidental scenario produces great intensity of thermal radiation during short periods.

This study demonstrates that the increase of the energy fluxes induces additional contributions, in comparison with simulations in the absence of the ground, to the dose and to IP index, that reach 137% and 167% for brick and concrete paving, respectively. Moreover, the safe distance - as specified in Italian civil protection limits - increases up to a maximum value of 50.4% for concrete.

Future studies will proceed with the computation of the fraction of reflected and re-emitted power striking the receptor. Likewise, the best conditions for admittance into the accident area, and the best intervention ways and times will be evaluated. Further accidental heat radiation scenarios, such as poolfire or other kinds of BLEVE/fireball will be studied through application of the the same simulation model. Finally, further studies will consider modification of the physical characteristics of materials with temperature, and the results derived will be compared to data from experimental tests.

Acknowlegements

The authors express their acknowledge to Dott. Maria Paola Bogliolo, ISPELS DIPIA, for helpful suggestions and contributions to this study.

References

Colwell R. N. (1983). *Manual of Remote Sensing* (Vol. 1). American Society of Photogrammetry, Falls Church, Virginia, USA.

DPC (1994). *Pianificazione di Emergenza Esterna per Impianti Industriali a Rischio di Incidente Rilevante* (Linee Guida), Dipartimento della Protezione Civile (1994).

Fletcher, C. A. J. (1991). *Computational Techniques for Fluid Dynamics*, Second Edition Vol.1, Springer Verlag.

Hunt, G. R. (1980). Electromagnetic radiation: the communication link in remote sensing. In Remote Sensing in Geology (Siegal and Gillespie, Eds) Chap. 2, pp 7–43. Wiley, New York,

Incropera F. P., De Witt D. P. (1996). *Introduction to Heat Transfer*, 3rd ed., John Wiley & Sons.

Lees, F. P. (1986). *Loss Prevention in the Process Industries*, Butterworth.

Nuovo Colombo, (1985). *Manuale dell'Ingegnere*, HOEPLI (81° Edizione) Milan.

Papazoglou, I.A. & Aneziris, O.N. (1999). Uncertainty qualification in the health consequences of the boiling liquid expanding vapour explosion phenomenon. *Journal of Hazardous Materials*, **A67**, 217-235.

TNO, (1992).*Methods for the calculation of physical effects of escape of dangerous materials* (Yellow Book), Committee for the Prevention of Disasters, Voorburg, Netherlands.

TNO, (1996) EFFECTS2 (1996). TNO Department of Industrial Safety, Holland.

TNO, (1983). *Hoofdgroep Maatschappelijke Technologie, in opdracht van het Ministerie van Volkshuisvesting*, Ruimtelijke Ordening en Milieubeheer. 'LPG-Integraal', vergelijkende risico-analyse van de opslag, de overslag, het vervoer en het gebruik van LPG en benzine; report no. 1112.

TNO, (1989). *Methods for determination of possible damage* (Green Book), Committee for the Prevention of Disasters, Voorburg, Netherlands.

8 Development and Application of an Extended Methodology to Validate Short-Range Atmospheric Dispersion Models

Harry Eleveld and Harry Slaper

National Institute of Public Health and the Environment (RIVM),
P.O. Box 1, 3720 BA Bilthoven, Netherlands

Abstract. The validation of real time atmospheric dispersion calculations requires an evaluation of the temporal and spatial characteristics of air concentrations and deposition. An extended statistical method is described for the validation and intercomparison of short-range atmospheric dispersion models. With this method multiple aspects of air dispersion modelling can be validated, including an evaluation of the spatial distribution. Ten statistical parameters are used and an overall ranking parameter is based on the combination of all ten parameters. The ranking parameter ranges from perfect agreement (value of 0) to extreme disagreement (value of 100)) and is constructed by scaling each of the statistical parameters on a scale from 0 to 100 and averaging the results for all parameters. In addition, nonstatistical parameters, which partly explain the calculated ranking, complement the extended method.
An application of the methodology illustrates its use in the comparison of calculated air concentrations from two dispersion models with measured air concentrations obtained in 19 days of dispersion experiments, which were available in the well-known Kincaid data set. The ranking parameter is found to vary considerably from day to day and between the models, from slightly under 20 to around 85. Bad results for one day for both models could be attributed to apparent deviations in the wind direction as given in the Kincaid data set.

8.1 Introduction

In the case of large-scale emergencies in nuclear or chemical industries real time atmospheric dispersion models can be an important tool to support decision-makers with spatial and temporal information on the dispersion and extent of the contamination. The real time calculation of short-range air dispersion is complicated due to the high variability of the relevant local atmospheric conditions. The accuracy of real time calculations of spatial and temporal distributions of the air concentrations is limited due to uncertainties in the applied dispersion algorithms and limitations in the available meteorological information. Thus, even for the most complicated algorithms, the agreement between modelled and observed concentrations will not be optimal. It is important to investigate the accuracy of the spatial and temporal distribution of modelled concentrations.

Using frequency distributions of air concentrations or crosswind integrated concentrations a good agreement between predictions and observations is found (Jaarsveld 1995). This implies that on average a model can provide adequate estimates, but this does not necessarily hold true for specific situations.

In order to validate real time atmospheric dispersion models, frequency distributions are applied in the Model Validation Kit (MVK) of Olesen 1996, which is based on the pioneering work of Hanna *et al.* (1991). Hanna *et al.* already included some basic statistical parameters in their BOOT software package. Cross wind integrated concentrations are applied in the ASTM90 methodology of Irwin (1999) to validate dispersion models. The use of either method, Olesen's or Irwin's, has some drawbacks.

1999, Cooper 1999, McHugh 1999, Venkatram 1999, Hantke 1999) and the developer Olesen (1999) also points out limitations of the MVK method. In its present form the method evaluates maximum concentrations on arcs at several distances from the source, and the comparison between modelled and observed concentrations is presented in so-called quantile-quantile (QQ) plots of non-corresponding predicted and observed data. A perfect match is obtained if the frequency distribution of the observed values equals the frequency distribution of the calculated concentrations. This implies that for a perfect model on average the maximum concentration on an arc is predicted well, but errors in the direction of the plume or errors in specific situations are not addressed in the evaluation.

The ASTM90 method, which is still under development, focuses on concentrations near the centre line of the plume and crosswind integrations of the plume concentrations. Olesen states in a recent paper (Olesen 1999) that the Kincaid data distributed with the ASTM90 method is inconsistent and that the method selects a 'perfect model' which predicts (too) low concentrations. In this paper an extended statistical methodology is presented to validate atmospheric dispersion models. The statistical parameters used are adapted from the European Tracer EXperiment (ETEX) (Mosca 1998), which was focused on the validation of long range atmospheric dispersion calculations. A ranking parameter is derived from the combined statistical parameters and can be used to compare the performance of the models for various situations. The application of the method is illustrated by using the Kincaid data set in an evaluation of two short-range air dispersion models (TADMOD and TSTEP). The presented evaluation of the models focuses on daily integrated air concentrations, a quantity which is relevant in the evaluation of the consequences of accidental releases from nuclear industries.

8.2   Model Validation Tool

The essential problem of performing deterministic calculations is shown in Figure 8.1.

With respect to the hour by hour values and specific receptor points the plot looks like a random distribution of points[1]. Another interesting point to be noted is the convergence of the data when daily-integrated concentrations are applied. The extended model validation method is basically taken from ETEX

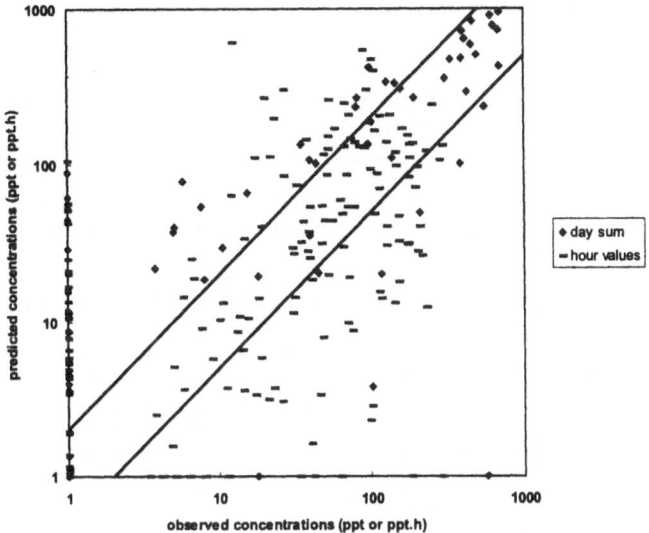

Fig. 8.1. Scatter plot of TADMOD for 31/5/81, day no. 19. The zero concentrations are raised to 1ppt (hourly averaged values) or ppt.h (day sums) in order to indicate the number of 'missing' data.

(Mosca 1998) and is complemented with non-statistical spatial information. Most of the statistical parameters, some in a slightly different formulation, are also included in the BOOT software of Hanna *et al.*. 1991.

The method uses ten different statistical parameters to identify the agreement or disagreement between modelled and measured (time-integrated) air concentrations. Among the parameters are: the mean differences and standard deviations between modelled and measured concentrations (arithmetic and geometric bias and standard deviations), the Pearson correlation coefficient, the Figure of Merit in Space (FMS) to illustrate the overlap between the modelled and measured cloud, the fraction of over and under predicted values, the number of predictions within a certain factor from the measured values.

The formulae for the parameters are described in Appendix 1. Using all separate statistical parameters, an overall ranking parameter is calculated as a

---

[1] It must be noted that the figure cannot be compared to the plots that can be generated with the MVK of Olesen. Those plots are based on arcwise maximum concentrations only; here we use results from all measured stations.

summarising quantity. A similar approach was adopted in the ETEX evaluation, but the ranking is redefined in a way that all separate parameters
are scaled from 0 to 100 points (0 – high quality, 100 – poor quality). For a
short description of the parameters see Table 8.1. The parameters are applied
to day-integrated air concentrations and will thereby overcome some of the
problems related to the chaotic nature of the atmosphere. The method can
also be applied to hourly values.

Table 8.1. Description of the applied parameters with respect to air dispersion
modelling (see Appendix 1 for detailed description)

Parameter	Description
BIAS	average difference between paired predictions and observations
NMSE'	the squared errors are pairwise (prediction vs observation) summed and normalised
geometric mean bias	model overestimation or underestimation is evaluated, the parameter is used for wide ranges of data
geometric mean variance	like NMSE, but in contrast to NMSE it gives the same weight to pairs showing the same ratio, also applied for wide ranges of data
Pearson's correlation coefficient	linear correlation of data
FA#	indicates fraction of data between factor # (2 or 5) over-predicting and underpredicting the observations
FOEX	index related to number overpredicted and underpredicted data
Kolmogorov Smirnov parameter	qualitative estimate of the difference between the cumulative distributions of the predicted and observed concentrations
Figure of Merit in Space	index related to the spatial overlap of the distributions in a way that receptor points are counted at which the predictions and observations are higher than a fraction of the maximum on grid
ranking	overall performance of the model

In order to acquire the statistical parameters the following steps were taken:

- the Kincaid data, see Appendix 2, as provided with MVK[2], are time
  integrated per day for each receptor point;

---

[2] in fact file: sf6_all.dat

- model output regarding daily integrated air concentrations is selected;
- a tool is written to interpolate with B-splines and cubic splines the predicted output grid to the Kincaid receptor points. In fact, the tool is suitable to interpolate any modelled grid to any observed data set, provided that the 'grids' do not differ too much;
- calculations of the actual statistical parameters.

In addition to the statistical parameters, the observed and predicted positions of the centre of mass are given as well as the angle between the lines connecting these points to the source.

The treatment of zeros, or very low values on the observation grid requires some consideration, because the statistical parameters can be sensitive to the treatment of those values. In a way, the number of zero values should not influence the agreement between model and measurement. In Appendix 3 the handling of these zeros, or low values, is explained.

The supposed benefit of using the above approach is that not only maximum values of modelled concentrations can be reviewed, but also

1. daily-integrated concentrations compare better than hourly averaged concentrations;
2. a spatial evaluation is possible;
3. a sensitivity/uncertainty study can be performed;
4. air dispersion models can be ranked.

8.3   Descriptions of Dispersion Models TADMOD and TSTEP

8.3.1   TADMOD

TADMOD is the name of the transport and deposition module of RASCAL. RASCAL version 1.3, which is developed for the US Nuclear Regulatory Commission (Athey *et al.* 1989), is implemented in the Radiological Emergency Management system (REM-3). This system, documented by Poley (1999), is developed to be used in the Dutch nuclear emergency response organisation and at the nuclear installations.

The REM-3 software consists of a model that describes the transport and deposition of radioactive nuclides after atmospheric emissions and models that assess the radiological dose consequences in terms of effective dose, red bone marrow dose, thyroid dose, lung dose and skin dose for adults, 10 and 1 year old children.

TADMOD uses a Lagrangian puff model to calculate the transport and dispersion of atmospheric released radionuclides. A unit emission at a constant rate during a certain time results in a number of discrete Gaussian puffs, which consist of the same amount of emitted material. The centres of the puffs are transported in the horizontal plane with the given wind vector. The mixing height and the ground are used as reflection layers for the calculation of the concentration. A description of a trigaussian puff model is used. The horizontal dispersion parameters (σ_x and σ_y) are assumed to be equal which simplifies the calculation of the concentration.

8.3.2   TSTEP

The model TSTEP has been developed as an interim model at RIVM for a first evaluation during nuclear emergencies. TSTEP is a descendant of the German model ATSTEP, which is one of the two short-range models that have been implemented in the European Decision Support System RODOS (c.f. e.g. Ehrhardt 1997). Major revisions are made in the ATSTEP algorithm for meteorology, plume transport and deposition; therefore the model TSTEP should be treated as a distinct model. No conclusion on the behaviour of ATSTEP should be drawn from TSTEP (or the other way around).

TSTEP is a segmented plume model; a continuous release is modeled as a time sequence of short, constant releases ('puffs'). During a short release (typically 5 to 15 minutes) stable meteorological conditions are assumed. A plume-shaped Gaussian distribution ('puff') describes the concentration profile. The plume widths are parameterized either by the height-dependent "Karlsruhe-Jülich" set or the uniform "Mol" set, as they are implemented in ATSTEP.

The horizontal plume spread ($\sigma_{x,y}$) is scaled with the observation time of the wind fields. The formulation of Briggs (1972) is used to calculate plume rise due to buoyancy. TSTEP assumes a linear rise as a function of source-distance, until the effective (final) height is reached. When a release is buoyant and its release height H_{stack} is larger than 80 % of the mixing layer height (0.8 $z_i < H_{\text{stack}}$), then TSTEP assumes that it is released *above* the mixing layer and will not disperse the puff.

8.4   Results and Discussion

The Kincaid dataset provides results of dispersion experiments for 24 days. Five of these experimental days were excluded from our evaluations because for those days the effective stack heights were above the mixing layer and both models used here are not capable of dispersion calculations above the mixing height. The results shown here are restricted to the 19 days where both models can perform dispersion calculations.

We will first show results for a few of the statistical parameters, focusing on the performance of both models for all selected days (Figures 8.2, 8.4, 8.6). The influence of the use of daily integrated or hourly averaged concentrations

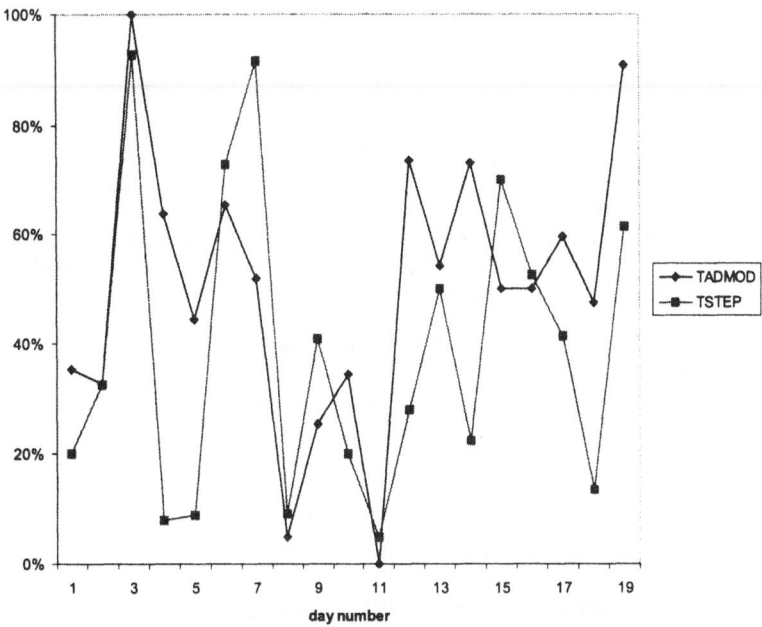

Fig. 8.2. FA5 for day sums for TADMOD and TSTEP. FA5 is based on the 100ppt.h cutoff (see text). High values indicate better agreement between predictions and observations.

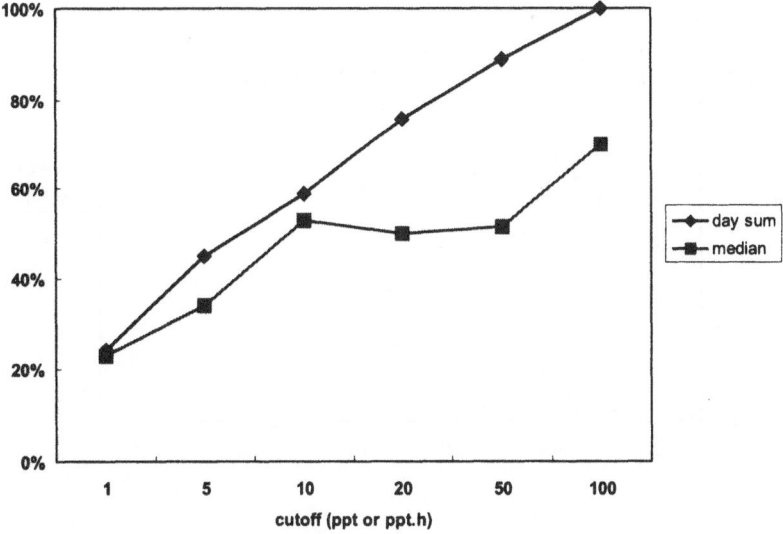

Fig. 8.3. FA5 for day sums and median day values for TADMOD for day 01−05−1980 (day no. 3) *vs* cutoff concentration. High values indicate better agreement between predictions and observations.

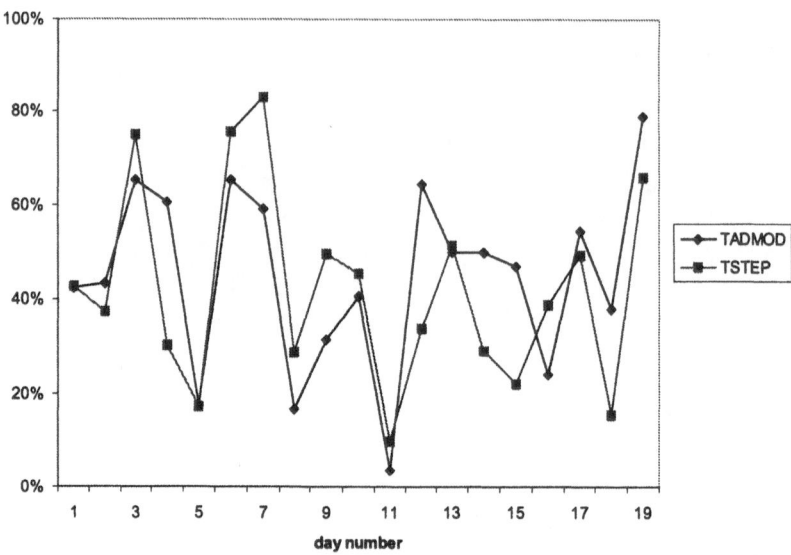

Fig. 8.4. Figure of Merit in Space (FMS) for day sums for TADMOD and TSTEP. The FMS is calculated for $0.1\times$ maximum grid value. High values indicate better agreement between predictions and observations.

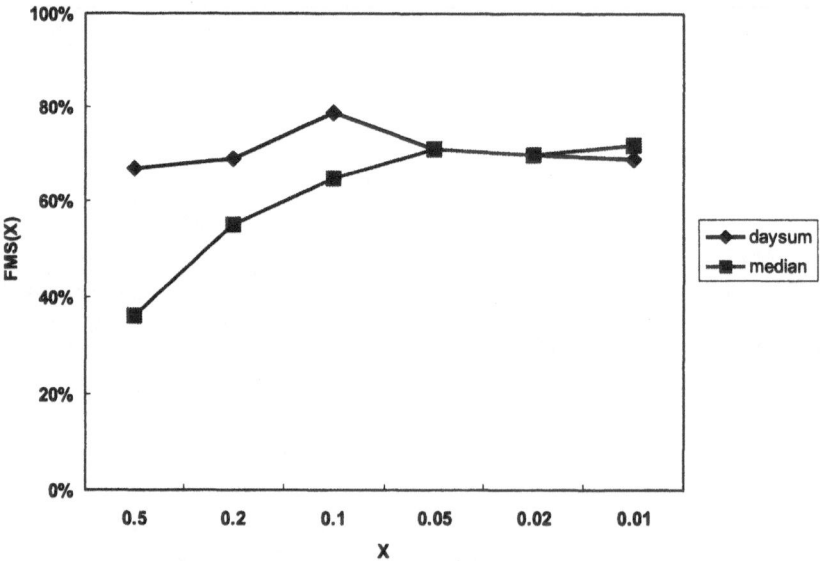

Fig. 8.5. FMS for day sums and median hour(ly averaged) concentrations for TAD-MOD for $31-05-1981$ (day no. 19). X is an indication for the fraction of the maximum value on the grid. High values indicate good agreement between predictions and observations.

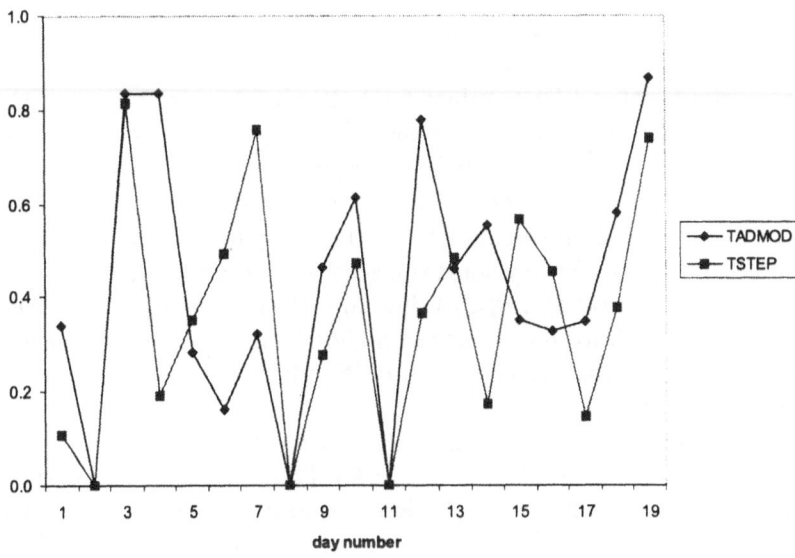

Fig. 8.6. Pearson's correlation coefficient for day sums for TADMOD and TSTEP. High values indicate good agreement between predictions and observations.

will be illustrated for some days in relation to choices with respect to the handling of low values (Figures 8.3 and 8.5).

For all 19 days the percentage of modelled data within a factor of 5 from the measured data (FA5) are shown in Figure 2. It is clear that the FA5 fluctuates between the best agreement (100 %) and no agreement (0%). A value of 100% for the FA5, indicating the best agreement, is converted to 0 for the overall ranking parameter as indicated in Appendix 1. The points where measured and modelled values are both below the cut off value of 100 ppt.h are excluded from the FA5 calculations. The average value amounts to 50% for TADMOD and 39% for TSTEP. Also, it is clear that FA5 for TADMOD is not always better than the FA5 for TSTEP. In 12 out of 19 cases the FA5 is higher for TADMOD. Furthermore, for days 8 and 11 the FA5 is very low for TADMOD. For days 3, 12, 14 and 19 the FA5 for TADMOD is rather high. The effect of setting specific cutoff values can be observed in Figure 8.3. By increasing the cutoff value the FA5 is increased. This result indicates that higher concentrations are predicted better. Also, the daily integrated concentrations give a higher FA5 than those based on median hourly averaged concentrations.

To illustrate the agreement on the position of the clouds the results for the calculation of the figure of merit in space are displayed in Figure 8.4. As for the FA5 the results for day 11 show a large discrepancy between both models on the one hand and the measurements on the other: the overlap (FMS) being less than 10% . Low values are also found for day 8 (in agreement with the low FA5 value) and day 5. On average the FMS-results for both models are comparable, but slightly better results are calculated for TADMOD (average FMS=45%) as compared to TSTEP (average FMS=42%). The day to day variation of the FMS values for both models show more correlation (Figure 8.4) than the variation for the FA5 (Figure 8.2). For days 3, 4, 12 and 19 the FMS is relatively high with respect to the TADMOD data.

The influence of the choice of the level (fraction of the maximum value on grid) at which the FMS is calculated is shown in Figure 8.5. For this particular day (no. 19) the FMS for day sums is insensitive for the level, while the FMS for the median hour(ly averaged) values are lower for higher levels. Also for other days the FMS for day sums are higher than those for median hour(ly averaged) values. A high value for the FMS indicates a good agreement with the observed data.

In Figure 8.6 Pearson's correlation coefficient for daily sums is shown for the two dispersion models. Again, the results for TADMOD are higher on the average than those for TSTEP: some 15%. Furthermore, TADMOD and TSTEP show zeros, i.e. no correlation, for days 2, 8 and 11. On the other hand, for days 3, 4, 12 and 19 the correlation coefficient is high with respect to TADMOD. A

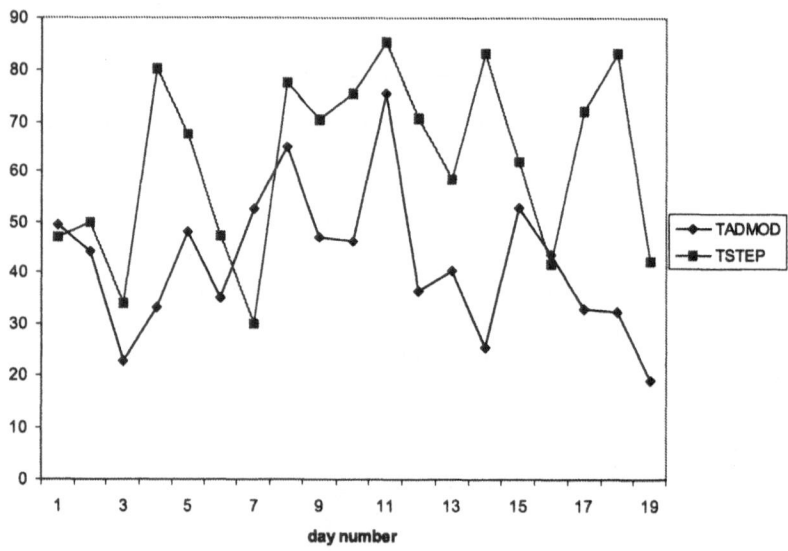

Fig. 8.7. Ranking for day sums for TADMOD and TSTEP. Low values indicate good agreement between predictions and observations.

summary of the overall result of the statistical comparisons is presented as the values for the ranking parameter in Figure 8.7. Good agreement between modelled and measured concentrations leads to low ranking values. The re-

sults for the TADMOD model are better than the ranking results for TSTEP
in 16 out of 19 days. According to the TADMOD ranking the performance of
TADMOD is poor for the days 7, 8, 15, and especially 11. The performance of
TADMOD is rather good according to the ranking for days 3, 14 and 19. We

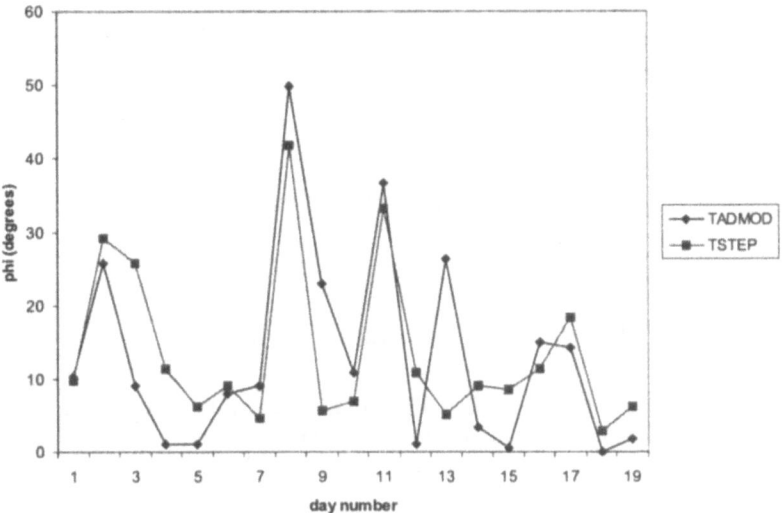

Fig. 8.8. Absolute angle between predicted centre of mass and measured centre of
mass.

will now look at certain non-statistical parameters to illustrate and substan-
tiate the findings in the overall ranking. We will focus on the position of the
centre of mass of the modelled and measured concentrations, by looking at
differences in the direction of the plume (Figure 8.8) and differences in the
distance from the source of the modelled and measured centres of mass of
the daily integrated concentrations (Figure 8.9).
To illustrate some of the aspects leading to high and low ranking values some
plots of observed and modelled dispersion are shown using a Geographical
Information System (Figures 8.10, 8.11, 8.12). The difference in the direction
of the plume between model and measurement is indicated by the error-angle,
which is defined as the angle between the directions of the measured and
modelled centres of mass as seen from the source. There is good agreement
between both models with respect to the error in the direction of the plume
(correlation coefficient 0.74 for the error-angle). For two occasions the error-
angle is more than 30 degrees, and Figure 8.10 illustrates the results for one
of these occasions (day 11). The wind direction as provided in the Kincaid
data set appears to be totally off, probably by as much as 90 degrees. The

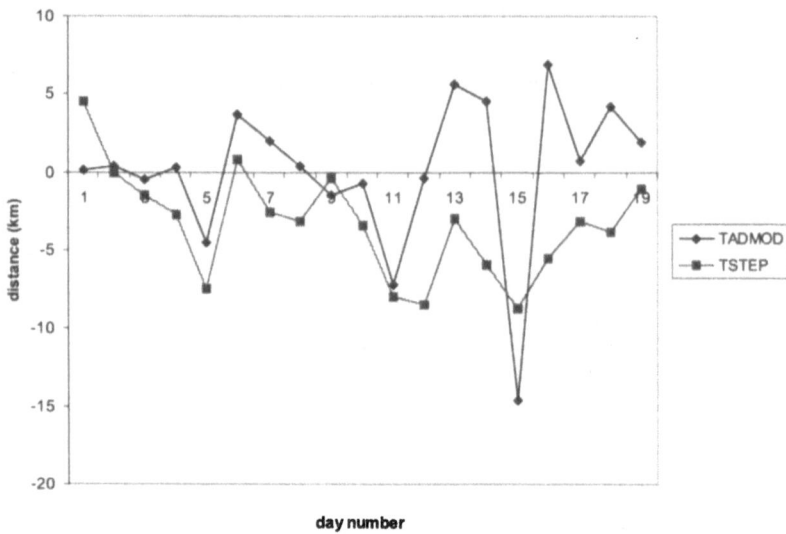

Fig. 8.9. Distance between predicted centre of mass and measured centre of mass. A negative distance means that the predicted centre of mass is closer to the source.

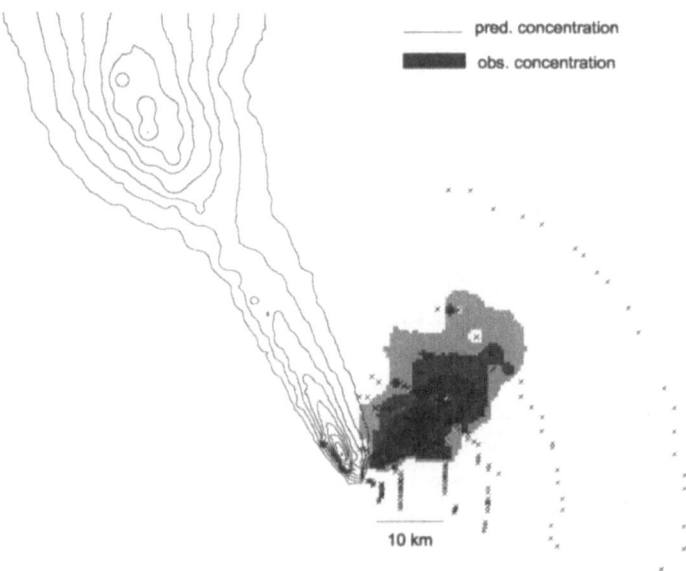

Fig. 8.10. Day sum of observed air concentrations and those predicted by TADMOD for day no. 11: 25 − 7 − 1980.

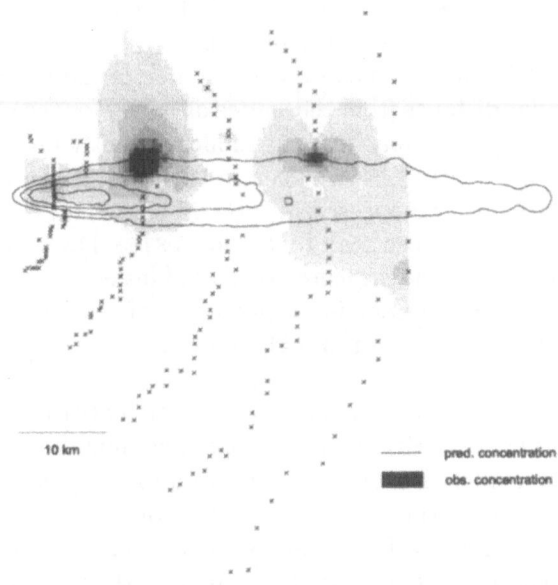

Fig. 8.11. Day sum of observed air concentrations and those predicted by TADMOD for day no. 5: 5 − 5 − 1980.

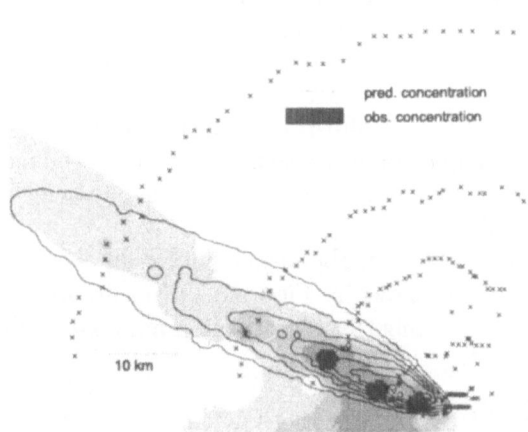

Fig. 8.12. Day sum of observed air concentrations and those predicted by TADMOD for day no. 12: 16 − 5 − 1981.

full error is not expressed in the error-angle because the calculated plume misses the observation posts (see Figure 8.10).

Figure 8.9 compares the distance from the source of the modelled centre of mass with a similar distance from the source obtained for the observed centre of mass. Negative values imply that the modelled centre of mass is closer to the release source than the observed centre of mass. As can be seen in Figure 8.9 the results for TADMOD scatter around zero, with on average +0.08 km distance, whereas the centres of mass for TSTEP have a tendency to be closer to the source than the measured centres of mass (average distance –3.3 km). It seems that the present implementation of TSTEP leads to ground level concentrations which are too high close to the source. This is further supported by the high percentage of overestimations that occur for TSTEP (FOEX[3] 25% for TSTEP, as compared to 1% for TADMOD) and is also reflected in various other statistical parameters (like small FA2 and FA5). This behaviour of the model also explains the overall higher ranking, indicating lower performance of TSTEP as compared to TADMOD.

We will now focus on a few days as an example to illustrate the relevance of the ranking values. The overall very bad results for the ranking for day 11 for both models is due to errors in the wind direction provided with the Kincaid data in the MVK (Figure 8.10). This result also illustrates the fact that the performance of a model in comparison with measurements is influenced not only by the model, but also by the quality of the input, e.g. the meteorological and release data. The bad ranking performance of the models for day 8 are also related to errors or at least uncertainties in the wind direction. The large, and thus bad, ranking for day 15 for TADMOD is not caused by the error in the wind direction, but relates to a large error in the calculated distance of the centre of mass from the source.

Day 5 also shows relatively bad results for the ranking, but the wind direction and the distance of the centre of mass are both in agreement with the measurements. Figure 8.11 illustrates the results for this day, showing the complexity of air dispersion in particular situations: the measured data show two separated high concentration areas for this day, which are not shown in the modelled results.

The geographical pattern for the dispersion for an average ranking value for the TADMOD model is shown in Figure 8.12 (results for day 12). This result shows quite good agreement in the direction and extension of the plume.

8.5   Conclusions

The extended methodology validates multiple aspects of the air dispersion modelling. Using this validation tool a spatial evaluation is possible, meanwhile preserving the capability of scoring the maximum predicted values on

---

[3] The FOEX varies from −50% to 50% , indicating total underprediction and total overprediction.

the grid as well. Furthermore, air dispersion models can be ranked when the validation tool is employed. The performance of the models can partly be explained by examining non-statistical parameters such as the differences in the directions and distances of the centres of mass of the predicted and observed 'puffs'.

By applying the validation tool it turns out that the daily-integrated concentrations compare better than hourly-averaged concentrations. The time integrated concentrations are applied to obtain quantities which are most relevant from the perspective of radiation protection.

The results from the ranking and validation procedure clearly demonstrate that the quality of the modelled values as compared to the observed values not only depend on the quality of the model, but also on the quality of the meteorological (and release) inputs. In addition, the validation methodology can be used as a powerful tool to evaluate the consistency of an observed data set. It was found that the input data for 25 July 1980 of the Kincaid data set, at least in the MVK, must be re-evaluated.

The methodology can be used to validate and intercompare models, and ranking results can be used to select models with the best overall performance.

The methodology described here could also be applied in other research areas where spatial and temporal dependent quantities for models are to be compared.

Appendix 1: Formulae for statistical parameters used

The formulae are constructed using the following variables:

N	number of receptor points,
P_i	predicted value at receptor point i,
M_i	observed value at receptor point i,
$\bar{P}$	grid-averaged predicted value,
$\bar{M}$	grid-averaged observed value,
$N_{(P_i > M_i)}$	number of receptor points at which $P_i > M_i$,
$N_{(M_i/\alpha < P_i < \alpha M_i)}$	number of receptor points at which $M_i/\alpha < P_i < \alpha M_i$,
N_{params}	number of statistical parameters used,

A_i — area of predicted ($i = 1$) or observed ($i = 2$) values for which the value of the concentration is above some specified threshold (10% of the maximum value on grid),

$\text{prob}(P(x_b))$, $\text{prob}(M(x_b))$ — probability of occurrence of predicted or observed values not higher than x_b

bias — $\frac{1}{N} \sum_i (P_i - M_i)$,

NMSE′ — $\sum_i (P_i - M_i)^2 / (P_i + M_i)^2$,

VG (geom mean variance) — $\exp\left\{ \frac{1}{N} \sum_i (\ln(M_i/P_i))^2 \right\}$,

MG (geom mean bias) — $\exp\left\{ \frac{1}{N} \sum_i \ln(M_i/P_i) \right\}$,

FOEX — $100 \left(N_{(P_i > M_i)} - 0.5N \right) / N$,

FMS — $(A_1 \cap A_2)/(A_1 \cup A_2)$,

KS (Kolomogorov Smirnov parameter) — $N \cdot \max |\text{prob}(P(x_b)) - \text{prob}(M(x_b))|$,

P_{corr} — $\sum_i (M_i - \bar{M})(P_i - \bar{P}) / \left(\sum_i (M_i - \bar{M})^2 \sum_i (P_i - \bar{P})^2 \right)^{1/2}$,

FAα — $N_{(M_i/\alpha < P_i < \alpha M_i)} / N$,

Ranking — $\frac{100}{N_{\text{params}}} \left\{ \frac{|\text{bias}|}{(\bar{P} + \bar{M})} + \text{NMSE}' + \frac{\ln(MG)}{|\ln(\bar{M}/\bar{P})|} + \frac{2|\text{FOEX}|}{100} \right.$

$\left. + (1 - \text{FMS}) + (1 - \text{FA2}) + (1 - \text{FA5}) + \frac{KS}{N} + (1 - P_{corr}) \right\}$

Note that NMSE′ is used instead of the standard

$$\text{NMSE} = \frac{1}{N} \sum_i (P_i - M_i)^2 / (\bar{P} \cdot \bar{M})$$

because of the possibility of scaling NMSE′ from 0 to 100.

Appendix 2: Day numbering and actual days in Kincaid data set

day number	date	day number	date
1	1980-04-20	11	1980-07-25
2	1980-04-25	12	1981-05-16
3	1980-05-01	13	1981-05-22
4	1980-05-04	14	1981-05-23
5	1980-05-05	15	1981-05-24
6	1980-05-07	16	1981-05-25
7	1980-05-09	17	1981-05-28
8	1980-07-11	18	1981-05-29
9	1980-07-13	19	1981-05-31
10	1980-07-24		

Appendix 3: The handling of zeros on the grids

Parameter	Handling of zeros
Bias	No special treatment
NMSE'	No special treatment
Geometric mean bias	Zeros are raised to a minimum value (1 ppt.h w.r.t. daily sums, equal to the detection limit at any hour) to overcome division by zero and to treat both predictions and observations in the same way
Geometric mean variance	Treated as geometric mean bias
Pearson correlation coefficient	No special treatment
FAα	In fact FA2 and FA5 deal with thresholds (T): if $P_i > T$ or $M_i > T$ then FAα is calculated
FOEX	Don't take the number of zeros (receptor points at which $P_i = M_i = 0$) into account, because a large grid will induce a low FOEX when zeros are used
Kolmogorov Smirnov parameter	No special treatment
FMS	No special treatment; the calculation is based on fractions of the maximum (so zeros are not of any importance)
Ranking	Follow from the above

References

Athey, G.F., Shoreen, A.L. & McKenna, T.J. (1989). Rascal Version 1.3, Users Guide . NUREG/CR-5247, ORNL/TM-10955, September 1989.

Briggs GA. (1972). Discussion: chimney plumes in neutral and stable surroundings. Atmospheric Environment, vol. **6**.

Cooper, N. S. (1999). A Review of the 'Model Validation Kit' (BOOT) and the draft ASTM validation procedures. Preprint of the Proceedings of the 6th International Conference on Harmonisation within Atmospheric Dispersion Modelling for Regulatory Purposes. October 11-14, 1999, Rouen, France.

Ehrhardt, J., Brown, J., French, S., Kelly, G.N., Mikkelsen, T. & Müller H. (1997). RODOS: Decision-making support for off-site emergency management after nuclear accidents, *Kerntechnik*, vol. **62**.

Hanna, S. R., Strimaitis, D. G. & Chang, J.C. (1991). *Hazard Response Modeling Uncertainty (a quantitive method) Volume I. User's guide for software for evaluating hazardous gas dispersion models.* Sigma Research Corporation. Westford, Ma (USA).

Hantke, T. & Eleveld, H. (1999). Validation of the Gaussian puff models TSTEP and REM3 using the Kincaid data set. Preprint of the Proceedings of the 6th International Conference on Harmonisation within Atmospheric Dispersion Modelling for Regulatory Purposes. October 11-14, 1999, Rouen, France.

Irwin, J. S. (1999). Effects of concentration fluctuations on statistical evaluation of centerline concentration estimates by atmospheric dispersion models. Preprint of the Proceedings of the 6th International Conference on Harmonisation within Atmospheric Dispersion Modelling for Regulatory Purposes. October 11-14, 1999, Rouen, France.

Jaarsveld, J. A. van (1995). *Modelling the long-term atmospheric behaviour of pollutants on various spatial scales.* PhD thesis. University of Utrecht, The Netherlands.

McHugh, C.A., Carruthers, D. J., Higson, H. & Dyster, S. J. (1999). Comparison of model evaluation methodologies with application to ADMS 3 and US models. Preprint of the Proceedings of the 6th International Conference on Harmonisation within Atmospheric Dsipersion Modelling for Regulatory Purposes. October 11-14, 1999, Rouen, France.

Mosca, S., Bianconi, R., Bellasio, R., Graziani, G. & Klug, W. (1998) ATMES II- Evaluation of long-range dispersion models using data of the 1st ETEX release. ISBN 92-828-3655-X. European Communities 1998. EUR 17756 EN.

Olesen, H. R. (1994). Model Validation Kit for the workshop on Operational Short-Range Atmospheric Dispersion Models for Environmental Impact Assessments in Europe. Mol, November 21-24 1994. June 1994. Reprint March 1996.

Olesen, H.R. (1999). Model Validation Kit – Recent developments. Preprint of the Proceedings of the 6th International Conference on Harmonisation within Atmospheric Dispersion Modelling for Regulatory Purposes. October 11-14, 1999, Rouen, France.

Poley, A.D. (1999). Documentatie Rekenmodellen REM-3, internal report NRG Petten no 21496/99.24076, in Dutch.

Venkatram, A. (1999). Applying a framework for evaluating the performance of air quality models. Preprint of the Proceedings of the 6th International Conference on Harmonisation within Atmospheric Dispersion Modelling for Regulatory Purposes. October 11-14, 1999, Rouen, France.

9 Uncertainty and Sensitivity of Dispersion Model Results to Meteorological Inputs: Two Case Studies

Joseph C. Chang

School of Computational Sciences, George Mason University,
4400 University Drive, Fairfax, VA 22030, USA. **

Abstract. Two case studies are presented to study the uncertainty and sensitivity of dispersion model results to meteorological inputs. Although the two case studies are distinctly different in terms of their scales and data availability, they are both common scenarios. The first case study mainly focuses on model sensitivity, and the second mainly focuses on model uncertainty.

The first case study involves the evaluation of the CALPUFF, HPAC, and VLSTRACK dispersion models with the Dipole Pride 26 (DP26) field data. DP26's test domain, ~30 km in scale, is a relatively flat dry lake bed surrounded by mountains. The experiments were well instrumented to measure the source, meteorology, and concentrations of the SF_6 tracer gas. The three dispersion models were first run in their 'standard' modes, where all models used the same raw meteorological data. Two additional HPAC sensitivity runs were made with variations in the choices of (i) the input raw meteorological data and (ii) the two diagnostic wind field models in HPAC. Overall, the performance of CALPUFF and HPAC (the base case) was comparable. VLSTRACK tended to overpredict and had a larger scatter. For example, for the maximum dosage along a sampling line, the fraction of predictions within a factor of two of observations was about 40% – 50% for CALPUFF and HPAC, and 20% for VLSTRACK. The results for the HPAC base run and the two sensitivity runs, although comparable, were often significantly different at the 95% confidence limits. A second set of sensitivity studies was also conducted where all dispersion models were driven by the same gridded wind fields prepared using the CALMET diagnostic wind model. It was found that the dispersion results were quite sensitive to the modeling options of CALMET.

The second case study involves the modeling of potential releases of chemical warfare agents during the 1991 Persian Gulf War. This case is different from DP26 in that the domain of interest was of the order of hundreds of kilometers, and that no onsite source, meteorological, and concentration data were collected. State-of-the-art prognostic mesoscale meteorological models (COAMPS, MM5, and OMEGA) were applied to simulate the prevailing meteorological conditions over Iraq. Wind fields generated by these models were then used by two dispersion models, HPAC and VLSTRACK. When compared to the available observations in countries surrounding Iraq, the three mesoscale models had errors in surface wind directions in the range

** also affiliated with TRW Systems Inc., assigned to the Office of the Special Assistant to the Under Secretary of Defense (Personnel & Readiness) for Gulf War Illnesses, Medical Readiness and Military Deployments, 5113 Leesburg Pike, Falls Church, VA 22041, USA.

of 30° to 90°. However, over Iraq where no observations were available for comparison, the uncertainty in predicted surface wind directions was fairly constant at about 70°. As a result, predicted dispersion patterns based on these mesoscale wind fields can sometimes be quite different. This disparity, although quite surprising and even disheartening, is, on reflection, to be expected. Moreover, it was found that the uncertainty in the global analyzed fields, used as the initial and boundary conditions for the mesoscale models, was already of the order of 30° to 60° for near-surface wind directions. It is unlikely that this uncertainty can be reduced by much because of random turbulence in the atmosphere.

9.1   Introduction

Dispersion models are powerful computer tools to predict the fate of pollutants after they are released into the atmosphere. The models account for the effects of atmospheric turbulent diffusion on diluting the pollutant source based on prevailing meteorological conditions (e.g. Arya 1999, Hanna *et al.* 1982). Pollutants can be contaminants routinely emitted by industrial sources (such as sulfur dioxide emissions from power plants), hazard chemicals released due to accidents (such as the rupture of a railroad car containing freon), or chemical and biological agents disseminated by weapon systems (such as sarin-filled rockets). It is imperative that these dispersion models be properly evaluated against observations before their predictions can be used with confidence. It is equally important to understand how sensitive model results are to input data and assumptions, and the degree of uncertainty in the results.

Dispersion is primarily controlled by turbulence in the atmospheric boundary layer. Turbulence is random by nature. The same two field experiments performed under the same atmospheric conditions will almost always yield different results. This leads to uncertainty in observations. On the other hand, due to factors such as errors in input data, model physics, and numerical representation, the model results also have uncertainty. According to Fox (1984), Anthes *et al.* (1989), and Hanna *et al.* (1991), uncertainty in dispersion modeling generally can be due to:

- Random turbulence:
  This mainly refers to the fluctuation in observed concentration (e.g. Csanady 1967, Chatwin and Sullivan 1979, and Wilson 1994). The uncertainty due to random turbulence cannot be reduced.
- Data errors:
  Models require data (e.g. observed meteorology) to run, and data (e.g. observed concentrations) to evaluate. These data can have errors due to reducible instrument errors (e.g. improper calibration), or irreducible instrument errors (e.g. inherent accuracy limit). It is sometimes necessary to first 'process' raw input data before they can be used by dispersion models. This processing is also a source of error. Toxicological data relate

dispersion modeling results to health effects and risk assessments. These data clearly carry a high degree of uncertainty.

- Model errors:
  These include errors in model physics and in translating from theory to numerical representation in the code. This type of error is not likely to be totally reducible.

To further demonstrate data errors, Figure 9.1 shows that the complete dispersion modeling process typically includes these four major steps: (i) the source term, land characteristics, and meteorology as dispersion model inputs; (ii) dispersion modeling; (iii) toxicology; and (iv) exposure assessments. This paper mainly focuses on the first two steps. Note that Figure 9.1 is

Typical dispersion modeling process

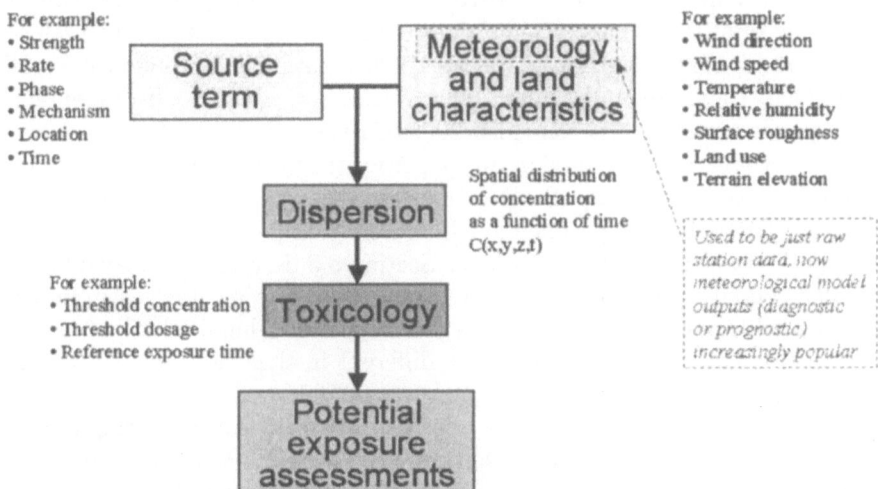

Fig. 9.1. Schematic of a typical dispersion process. Note that each process has certain degrees of uncertainty.

oversimplified in a sense that meteorological modeling is by itself perhaps an even larger research area than dispersion modeling. It is now increasingly popular to use sophisticated mesoscale models to generate the necessary high-resolution meteorological fields for dispersion modeling, especially over areas where measurements are sparse (e.g. Anthes *et al.* 1989). This introduces additional sources of uncertainty in the final dispersion results. For example, the mesoscale model outputs will depend on (i) how various physical processes are parameterized, (ii) how surface characteristics are specified, (iii) how models are initialized with large-scale fields, and (iv) whether observations are dynamically assimilated during the course of numerical integration. The decisions on all of these factors will ultimately impact on the location, shape, and size of predicted plume footprints, which in turn are translated into important measures such as the casualty rate. Consequently, it is crucial to consider the additional uncertainty and sensitivity as introduced by mesoscale meteorological modeling.

Different methods can be used to account for uncertainty due to random turbulence and data errors. Estimating the uncertainty due to random turbulence typically boils down to the determination of the moments (e.g. mean and variance) of the concentration time series, and the assumption of a corresponding probability distribution function (pdf). Concentration fluctuations can be estimated by at least these five types of methods: Lagrangian particle models (e.g. Sawford 1983, and Kaplan and Dinar 1993), large eddy simulations (e.g. Sykes and Henn 1992), turbulence closure models (e.g. Sykes *et al.* 1984), meandering plume models (e.g. Gifford 1959), and laboratory experiments (Willis and Deardorff 1976, 1978). The following methods can be used to account for the uncertainty due to data errors: Monte Carlo analysis (Hanna *et al.* 1998, 2000; Bergin *et al.*. 1999), ensemble methods (Stensrud *et al.* 2000; Debberdt and Miller 2000; Straume *et al.* 1998), and emulator (or response surface) methods (Fish 2000; Carslaw *et al.* 1999).

Two case studies are presented in this paper for the purpose of studying model uncertainty and sensitivity. Section 9.2 describes the first case, the Dipole Pride 26 (DP26) field experiments, which were well instrumented over a test domain that was about 30 km in scale. Section 9.3 describes the second case, the 1991 Persian Gulf War exposure study, where no onsite measurements were available and the domain of interest was several hundred kilometers in scale. These two cases, although quite different in their attributes, are both typical scenarios. A systematic approach was not followed here to study model uncertainty and sensitivity. Rather, representative ways of preparing meteorological inputs for dispersion models are presented in order to demonstrate model sensitivity, and how model uncertainty can be traced. Section 9.4 gives the conclusions.

9.2   Model Evaluation and Sensitivity Study with Dipole Pride 26

9.2.1   Descriptions of Dipole Pride 26

The Dipole Pride 26 (DP26) field experiments were conducted in November, 1996, at Yucca Flat (37°N, 116°W), Nevada Test Site, Nevada. The experiments were sponsored by the Defense Threat Reduction Agency (DTRA), with a primary goal to validate dispersion models. Watson *et al.* (1998) and Biltoft (1998) provide a detailed description of the experiments. Figure 9.2 shows the test site and instrument layout.

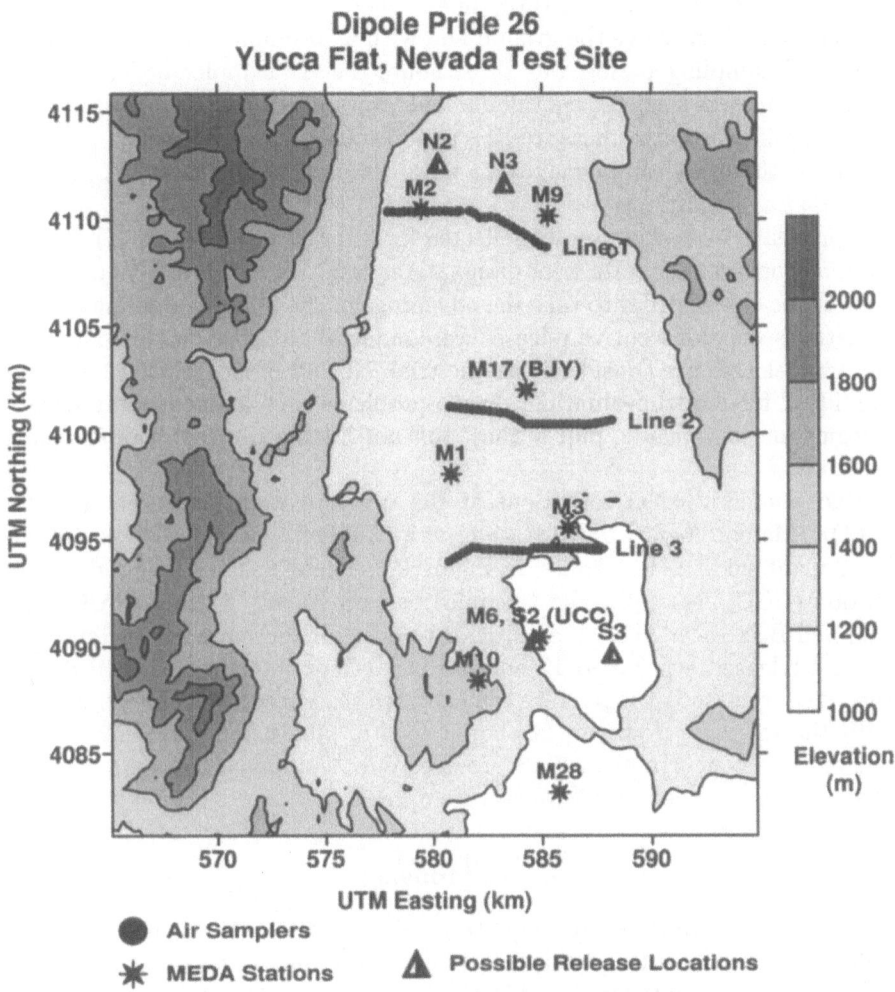

Fig. 9.2. The Dipole Pride 26 test site at Yucca Flat (37°N, 116°W), Nevada
Test Site, Nevada. Also shown are the three SF$_6$ sampler lines (small circles, 30
samplers per line), eight MEDA surface meteorological stations (stars), and four
possible release locations (triangles). There are also two pibal stations (BJY, near
M17; and UCC, near M6), and one radiosonde station (UCC, near M6). Terrain
resolution is 250m. The map covers an area of 30 × 35 km. The coordinates of the
southwest corner of the map are $(36.8745°N, 116.2707°W)$.

The experiments involved instantaneous releases of SF_6 at roughly 6 m above the ground. Depending on the prevailing wind direction at the test site, a release was either from the north (locations N2 and N3 in Figure 9.2) or from the south (locations S2 and S3 in Figure 9.2) of the test site. The main sampling array consisted of three lines, where each line had 30 whole air samplers at 1.5 m above the ground and with a 15-minute sampling interval. The total sampling period was three hours (i.e. 12 sampler bags) for each trial. There were also six continuous analyzers employed along the middle sampling line that can measure SF_6 concentrations at a frequency of 4 Hz. However, these high-frequency data were not used in this study because of poor spatial resolution.

Sampler data were successfully collected for a total of 21 releases. There are, however, only 14 separate trial designations (e.g. DSWA01, DSWA03, etc.). This is because in order to take the advantage of the 3-hour sampling period, sometimes two consecutive releases were made about 90 minutes apart, and the two releases are considered as one trial. Not all of the above 21 releases are 'ideal' for model evaluation, due to problems such as incomplete meteorological measurements, puff lofting, puff not hitting the center of sampling lines, etc.

Surface meteorological conditions at the test site were measured by eight MEDA (Meteorological Data) stations at a 15-minute interval. These stations are designated as M1, M2, etc. in Figure 9.2. There was also one radiosonde station (UCC, near M6), and two pilot balloon (pibal) stations (UCC, near M6; and BJY, near M17) that provided upper air measurements. Radiosondes are released every 3 to 12 hours. (The 3-hour interval was used during the tracer releases.) Pibals were typically released every hour in conjunction with the releases. However, at the UCC site, pibals and radiosondes were not released concurrently. Pibals provided only wind measurements, whereas radiosondes provided also temperature data.

9.2.2   Overview of Dispersion Models

The CALPUFF (version 5.0; Scire *et al.* 2000a), HPAC (version 3.2; DTRA 1999), and VLSTRACK (version 3.0; Bauer and Gibbs 1998) dispersion models have been evaluated with the DP26 data. CALPUFF has been traditionally used in Environmental Protection Agency's (EPA) regulatory applications, where environmental impacts due to routine industrial releases are modeled. HPAC and VLSTRACK, on the other hand, are models widely used by the US Department of Defense (DoD) to estimate potential hazards due to dissemination of chemical, biological, and nuclear weapons.

Fundamentally, all three models are based on a Gaussian puff dispersion formulation. However, the three models are different in several ways, such as their boundary layer parameterizations, their treatments of terrain and the transport wind field, their handling of land surface characteristics, and their data ingestion methods and requirements. CALPUFF has a built-in diagnostic meteorological model, CALMET (Scire *et al.* 2000b). HPAC has two built-in diagnostic meteorological models, SWIFT and MC-SCIPUFF. VLSTRACK, on the other hand, does not have a built-in wind field model, and has only a simple interpolation algorithms.

Although CALPUFF is a 'puff' model, it accepts only hourly average emission and meteorological information, and predicts only hourly average concentrations. As a result, the model cannot directly compare to high-frequency data that are typical of puff experiments such as DP26. This limitation primarily results from the fact that the model has been traditionally used in regulatory applications where hourly is the predominant time interval of choice. HPAC and VLSTRACK can readily accept and produce higher frequency data. Because all models need to be compared on an equal basis, DP26's higher-frequency data were mostly not used.

HPAC has the most flexible input data requirements among the three dispersion models. Moreover, the user can readily select one of the two integrated diagnostic models to prepare for mass-consistent wind fields. It is thus relatively easy to investigate with HPAC the effects of using different combinations of input data. In addition to the base case, two additional HPAC sensitivity runs were also considered in this study. Variations among the three HPAC runs were mainly in the selection of the wind field model and the raw meteorological data as summarized below:

	Wind field model	Meteorological observations
Base Run	SWIFT	Surface + upper air
Sensitivity Run 1	MC-SCIPUFF	Surface + upper air
Sensitivity Run 2	MC-SCIPUFF	Surface only

For the above three HPAC configurations, Base Run appears to be the most optimal, logical configuration since it uses (i) the more advanced SWIFT wind field model and (ii) more observed meteorology. Conversely, Sensitivity Run 2 appears to be the least sophisticated configuration. One way to determine model sensitivity is through performance evaluation. In the following, the performance of the HPAC base run is primarily used to compare against CALPUFF and VLSTRACK. The three HPAC runs are then evaluated as a group. The next section briefly describes the evaluation methodology.

Since CALMET requires temperature profiles, which pibals lacked, the pibal data were not used altogether for the sake of consistency. Furthermore, the radiosonde data near the surface were not used in the modeling because of questionable quality (Chang *et al.* 1999).

9.2.3   Evaluation Methodology

SF$_6$ concentrations were mainly measured by 90 whole air samplers with a 15-minute sampling period and a 3-hour total duration. These 15-minute concentrations were summed to derive the total dosage (ppt-hr) measured at each sampler, where a background SF_6 concentration of 5 ppt was subtracted. Model evaluation was mainly based on the maximum dosage anywhere along a sampling line. Model evaluation based on the summation of dosages over all samplers along a sampling line (a surrogate for the familiar cross-wind integrated concentration often considered by dispersion modelers) was also considered. Those results can be found in Chang et al. (1999) and are not presented here.

In order to evaluate how well model predictions compare against observations, the following statistical performance measures were considered (Hanna et al. 1993), including the fractional bias (FB), geometric mean bias (MG), normalized mean square error (NMSE), geometric mean variance (VG), and fraction of predictions within a factor of two of observations (FAC2), where

$$FB = \frac{(\overline{C_o} - \overline{C_p})}{0.5\,(\overline{C_o} + \overline{C_p})} \tag{9.1}$$

$$MG = \exp\left(\overline{\ln C_o} - \overline{\ln C_p}\right) \tag{9.2}$$

$$NMSE = \frac{\overline{(C_o - C_p)^2}}{\overline{C_o}\,\overline{C_p}} \tag{9.3}$$

$$VG = \exp\left[\overline{(\ln C_o - \ln C_p)^2}\right] \tag{9.4}$$

$$FAC2 = \text{fraction of data for which } 0.5 \le C_p/C_o \le 2.0 \tag{9.5}$$

where

C_p	denotes	predicted value,
C_o	denotes	observed value,
$\overline{C}$	denotes	average.

FB and MG are similar statistics measuring the systematic bias of a model. The former is based on a linear scale and the latter is based on a logarithmic scale. Likewise, NMSE and VG are similar statistics measuring the random scatter of a model, with the former based on a linear scale and the latter based on a logarithmic scale. A perfect model would have MG, VG and $FAC2 = 1.0$; and FB and $NMSE = 0.0$. A factor-of-two mean bias for C_p would mean that $FB = \pm 2/3$, $MG = 0.5$ or 2.0, $NMSE = 0.5$, and $VG = 1.6$.

Each performance measure has its advantages and shortcomings. The relative advantages of a measure are partly determined by the distribution of the variables of interest. FB and NMSE are more strongly influenced by infrequently occurring high observed and predicted concentrations, whereas MG and VG provide a more balanced treatment of both high and low values. For a dataset where both predicted and observed concentrations vary by many orders of magnitude, MG and VG would be more appropriate. However, VG is also known to be strongly influenced by extremely low values. This can occur when model predictions (C_p) are zero but observations (C_o) are finite. A small lower limit is often assigned to C_p/C_o so that it is possible to calculate VG. It is important to adopt a balanced approach to base the conclusion on a suite of performance measures.

Bootstrap resampling (Efron 1987) was used to estimate the mean, μ and standard deviation, σ of the above performance measures. The 95% confidence limits can then be defined as

$$\mu \pm t_{95\%}\sigma(n/(n-1))^{1/2} \qquad (9.6)$$

where n is the number of resamples, and $t_{95\%}$ is the two-sided 95% point of the Student t distribution.

9.2.4   Evaluation Results

Figure 9.3 shows the scatter plots of observed versus predicted maximum dosage (ppt-hr) for CALPUFF, the three HPAC runs, and VLSTRACK, where any values smaller than 1.0 were arbitrarily set to 1.0. Recall that out of the three ways of running HPAC, the base run is supposed to be the optimal configuration, and Sensitivity Run 2 should be the least sophisticated configuration.

CALPUFF and the three HPAC runs generally had good agreement with observations. CALPUFF seemed to overpredict slightly. VLSTRACK overpredicted the most. There were differences among the three HPAC runs. It is interesting to note that Sensitivity Run 2 with the least sophisticated configuration still had a relatively good performance. This is somewhat counter-intuitive.

Near-surface radiosonde measurements were ignored due to questionable quality. However, there were still noticeable differences between HPAC Sensitivity Runs 1 (with surface and upper air observations) and 2 (with surface observations only). This is due to the fact that for Sensitivity Run 2, MC-SCIPUFF assumed the winds are vertically uniform, whereas for Sensitivity Run 1, MC-SCIPUFF linearly interpolated between surface and upper air winds. The approach used in CALMET was more robust, where surface winds were vertically extrapolated, based on similarity theories, to blend with upper air data. In this case, surface winds carried more weights in the lower layers as opposed to the traditional scheme of straight linear interpolation.

All HPAC runs predicted some very low dosages, as indicated by the minimum value of 1.0 in Figure 9.3. Interestingly, very low dosages were mainly found in the HPAC base case and Sensitivity Run 1 where radiosonde data were used. CALPUFF did not have any low outliers, probably a result of the more robust wind field. There were six points located at the bottom axis for VLSTRACK in

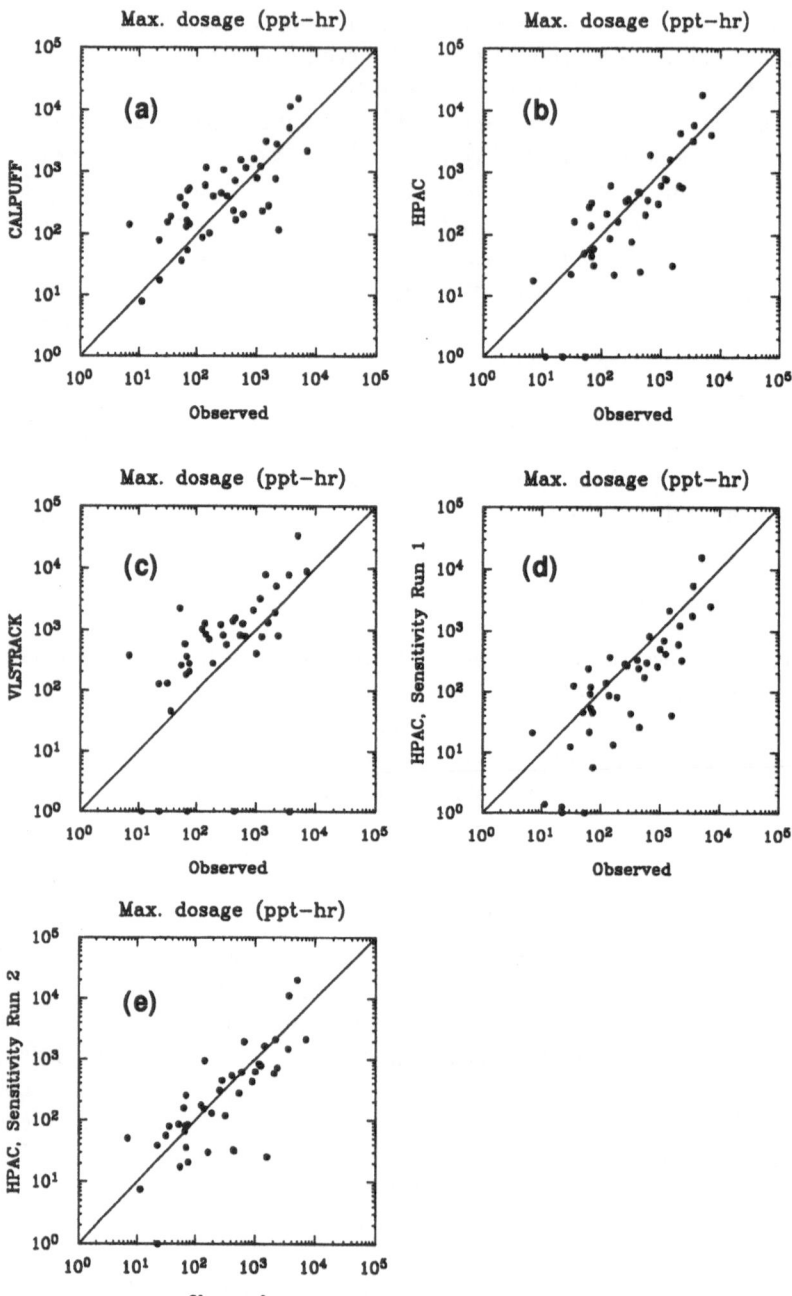

Fig. 9.3. Scatter plots of the maximum dosage (ppt-hr) at each sampling line for (a) CALPUFF, (b) HPAC base run, (c) VLSTRACK, (d) HPAC Sensitivity Run 1, and (e) HPAC Sensitivity Run 2. Any predicted value less than 1.0 (including zero) is set to 1.0.

Figure 9.3. This probably had more to do with the fact that VLSTRACK did not print any results when the predicted concentration was less than about 22 ppt (0.0001 mg/m^3).

Table 9.1 summarizes the five statistical performance measures described in (9.1) – (9.5), plus the average, standard deviation (S.D.), the highest value ($HI1$) and the second highest value ($HI2$) for the maximum dosage anywhere along a sampling line for CALPUFF, the HPAC base run, and VLSTRACK. The

Table 9.1. Summary of Performance Measures, Including MG (Geometric Mean Bias (9.2)), VG (Geometric Mean Variance (9.4)), FB (Fractional Bias (9.1)), $NMSE$ (Normalized Mean Square Error (9.3)), and FAC2 (Fraction of predictions within a factor of 2 of observations (9.5)) for the Maximum Dosage (ppt-hr) anywhere along a sampling line for CALPUFF, HPAC Base Run, and VLSTRACK. The Average, Standard Deviation (S.D.), Highest Value (HI), and Second Highest Value (HI2) of the data are also shown. A perfect model would have MG, VG, R, and $FAC2$ = 1.0; and FB and $NMSE$ = 0.0.

	Observed	CALPUFF	HPAC Base Run	VLSTRACK
Average (ppt-hr)	914	1336	1125	2150
S.D. (ppt-hr)	1463	2897	2929	5253
MG	n/a	0.670	1.607	0.764
VG	n/a	4.92	9.22	298
FB	n/a	-0.375	-0.207	-0.807
$NMSE$	n/a	4.10	4.60	10.75
$FAC2$	n/a	0.405	0.500	0.214
HI1 (ppt-hr)	7022	15340	18006	32963
HI2 (ppt-hr)	4987	11369	5810	8871

values of MG were between 0.670 and 1.607, which amounted to 50% overprediction to 40% underprediction based on a logarithmic scale. The values of VG were roughly equivalent to a factor of 4 to 10 random scatter based on a logarithmic scale. CALPUFF had the smallest VG. Although VLSTRACK had an MG that is closest to 1.0, Figure 9.3 shows that it is due to an overall overprediction balanced by a smaller number of extremely low predictions. The values of FB were between –0.207 and –0.807, which led to 25% to a factor of 2.4 overprediction based on a linear scale. The values of $NMSE$ were between 4.10 and 10.75, which were equivalent to about a factor 6 to 13 scatter based on a linear scale. HPAC had the smallest FB, and CALPUFF had the smallest $NMSE$.

Notice the much higher value of VG for VLSTRACK in Table 9.1. It is partly due to a systematic bias of overprediction (see Figure 9.3, also indicated by the value of FB), and partly due to the fact that VLSTRACK predicted no impact for a few cases where finite observational values were measured.

A small, arbitrary model residual, $C_p/C_o = 0.01$, was assumed in order to calculate the VG statistics. $NMSE$, on the other hand, does not have such a shortcoming.

While the fraction of predictions of the maximum dosage within a factor of two of observations was about 40 to 50% for CALPUFF and HPAC, it was only about 20% for VLSTRACK. All models overpredicted the highest and second highest dosages by roughly a factor of two to four.

Table 9.2 summarizes the statistical performance measures for the HPAC base case and the two HPAC sensitivity runs. The HPAC base run, suppos-

Table 9.2. Summary of Performance Measures, Including MG (Geometric Mean Bias (9.2)), VG (Geometric Mean Variance (9.4)), FB (Fractional Bias (9.1)), $NMSE$ (Normalized Mean Square Error (9.3)), and $FAC2$ (Fraction of predictions within a factor of 2 of observations (9.5)) for the Maximum Dosage (ppt-hr) anywhere along a sampling line for the HPAC Base Run and Two Sensitivity Runs. The Average, Standard Deviation (S.D.), Highest Value (HI), and Second Highest Value (HI2) of the data are also shown. A perfect model would have MG, VG, R, and $FAC2 =$ 1.0; and FB and $NMSE = 0.0$.

	Observed	HPAC Sensitivity Run 1	HPAC Sensitivity Run 2	HPAC Base Run
Average (ppt-hr)	914	839	1178	1125
S.D. (ppt-hr)	1463	2465	3429	2929
MG	n/a	2.235	1.337	1.607
VG	n/a	10.97	5.46	9.22
FB	n/a	0.086	-0.253	-0.207
$NMSE$	n/a	4.56	7.17	4.60
$FAC2$	n/a	0.429	0.500	0.500
HI1 (ppt-hr)	7022	15401	20142	18006
HI2 (ppt-hr)	4987	5414	11115	5810

edly the best configuration with the more advanced SWIFT wind model and more observed meteorology, did not necessarily lead to the best performance. Likewise, Sensitivity Run 2, supposedly the least sophisticated configuration with the MC-SCIPUFF wind model and less observed meteorology, did not necessarily result in the worst performance. For example, the values of MG, FB, and $FAC2$ for the base run and Sensitivity Run 2 of HPAC were quite comparable. The performance measures for the three HPAC runs were often significantly different at the 95% confidence limits.

9.2.5   Second Set of Sensitivity Studies

The above results were based on 'typical' ways of running the dispersion models. That is, each dispersion model was allowed to prepare gridded meteorological fields from raw data with its own meteorological models (for CALPUFF and HPAC) or a simple algorithm (for VLSTRACK). It was shown that even the three typical ways of running HPAC (the base run plus the two sensitivity runs) generated results that were often significantly different. This shows the sensitivity of dispersion model results to input meteorological fields, and suggests the need of further investigating model sensitivity to other options of specifying meteorological inputs. Since the CALMET diagnostic wind field model (Scire *et al.* 2000b) in CALPUFF is a separate program, and has its own input file whereby the user can easily specify various options, another set of sensitivity study is considered where the three dispersion models are driven by common CALMET wind fields. (The SWIFT and MC-SCIPUFF wind-field models are integral components of HPAC, and cannot be easily configured by the user.)

As mentioned above, one of the reasons for CALPUFF's more robust results (i.e. no low outliers in predicted concentrations) is that CALMET can vertically extrapolate surface wind observations, based on similarity theories, to supplement upper-air measurements. It is interesting to find out whether the CALMET wind field has the same effects on HPAC and VLSTRAC, and whether other CALMET wind fields created using different options will substantially affect predicted tracer concentrations. One important distinction of the current set of sensitivity studies from the earlier set is that instead of using the same *raw* meteorological data, all dispersion models are now using the same *gridded* meteorological data. In the following, three options of running CALMET will be considered. Most discussions will be qualitative, with the main emphasis to further demonstrate the sensitive dependence of the dispersion model results on the choice of the input meteorological fields.

The surface wind fields observed by the eight MEDA (Meteorological Data) stations over the DP26 test domain often showed large spatial variability due to surrounding mountains. Figures 9.4(a) and (b) show the observed surface wind fields for 0400 LST, November 8, 1996; and 0500 LST, November 11, 1996, respectively. In Figure 9.4(a), about half of the surface wind speeds were in the range of 5 m/s, and the remaining half were around 1 m/s. In Figure 9.4(b), while most MEDA stations measured moderate northwesterly winds, large discrepancy existed between stations MEDA06 (0.3 m/s and 60°) and MEDA03 (7.7 m/s and 0°). (See Figure 9.4(c) for station locations.) Diagnostic wind-field models such as CALMET will have to accommodate this large spatial variability, through different interpolating algorithms, to create gridded wind fields. Figure 9.5(a) shows the horizontal CALMET wind field for the surface layer (between 0 and 20 m above the ground).

The gridded wind field clearly adjusted to the observed wind field as shown in Figure 9.4. Figure 9.5(b) shows the horizontal CALMET wind field for the top layer (between 2000 and 3300 m above the ground). Spatial variability is still prominent because of the vertical extrapolation done by CALMET of surface wind observations. Figures 9.5(c) and (d) show the predicted verti-

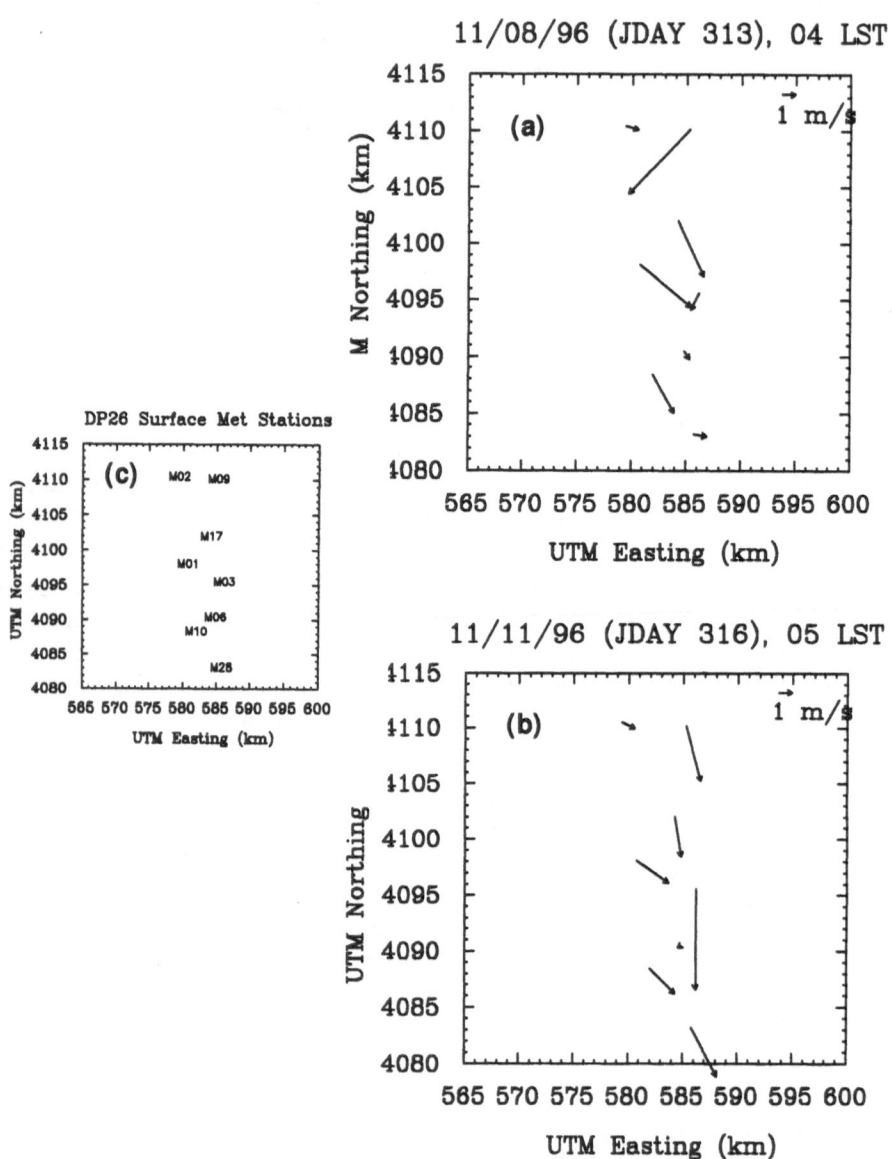

Fig. 9.4. Observed DP26 surface wind fields for (a) 0400 LST, Nov 8, 1996; and (b) 0500 LST, Nov 11, 1996. The names of the surface MEDA stations are shown in (c).

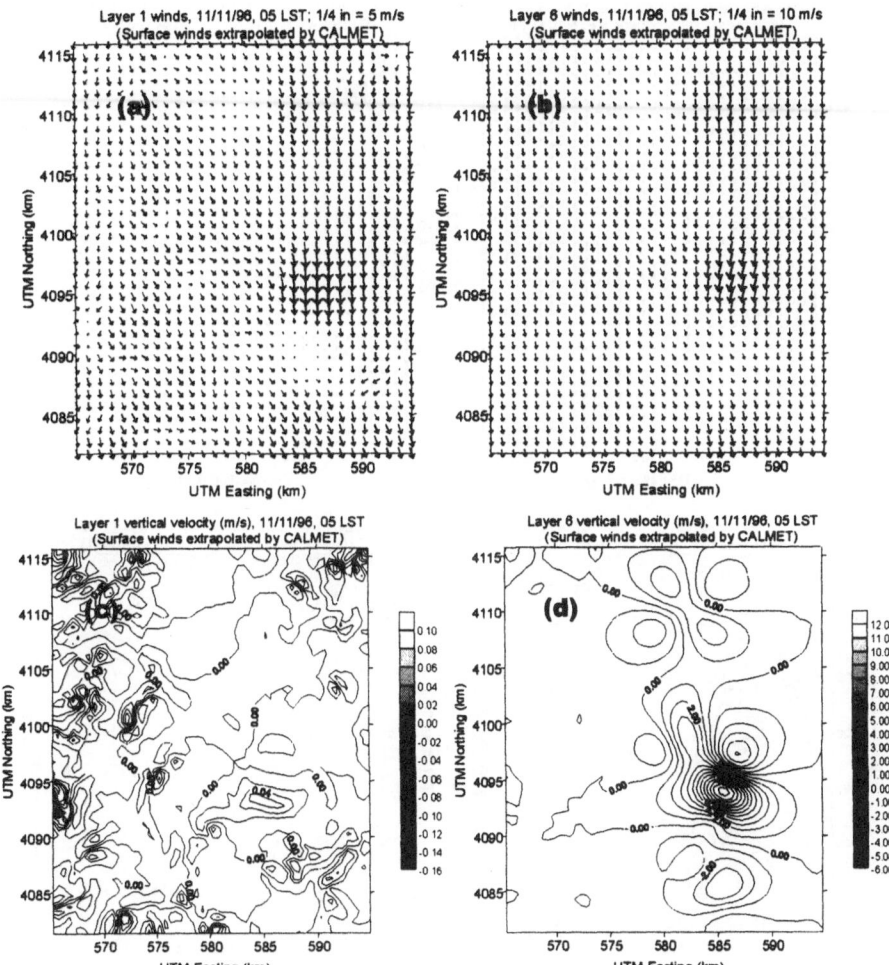

Fig. 9.5. Wind fields simulated by CALMET for 0500 LST, Nov 11, 1996, where surface wind observations were vertically extrapolated. (a) Horizontal wind vectors for Layer 1 (0 to 20 m above the ground). (b) Horizontal wind vectors for Layer 6 (2000 to 3300 m above ground). (c) Vertical velocity field for Layer 1. (d)Vertical velocity field for Layer 6. Note different scales in (c) and (d).

cal velocity fields for the surface and top layers. While the vertical velocity field for the surface layer was in the order of 0.1 m/s, the extrapolation of surface wind observations led to unreasonably large values for the vertical velocity, in the order of 10 m/s, over the region where surface wind observations showed the largest disparity. Figures 9.6(a) through (d) are the same as Figures 9.5(a) through (d), except that the vertical extrapolation option was disabled. This resulted in a uniform horizontal wind field in the top layer as shown in Figure 9.6(b), because there was only one upper-air measurement site; and relatively small vertical velocity, less than 0.5 m/s, in the top

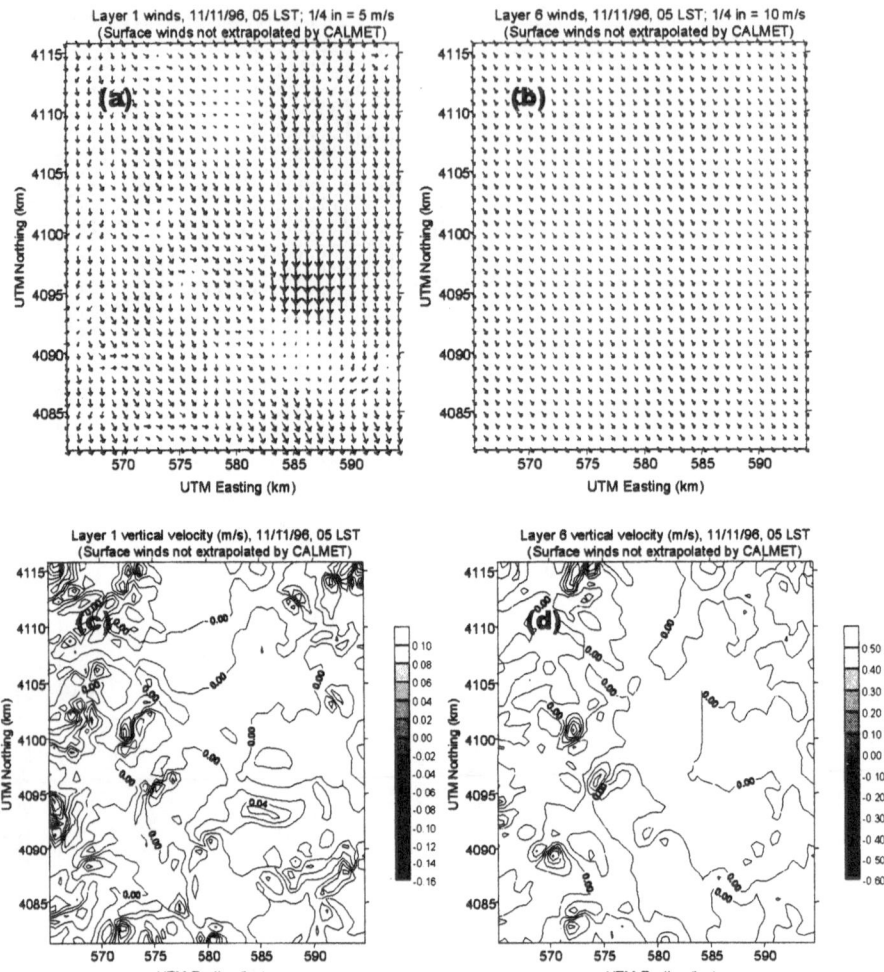

Fig. 9.6. Wind fields simulated by CALMET for 0500 LST, Nov 11, 1996, where surface wind observations were *not* vertically extrapolated. (a) Horizontal wind vectors for Layer 1 (0 to 20 m above ground). (b) Horizontal wind vectors for Layer 6 (2000 to 3300 m above ground). (c) Vertical velocity field for Layer 1. (d) Vertical velocity field for Layer 6. Note different scales in (c) and (d).

layer as shown in Figure 9.6(d). Comparison of Figures 9.5(a)-(b) with Figures 9.6(a)-(b) reveals that the surface fields are not affected by the disabling of the vertical extrapolation option.

Because of the unrealistic large vertical velocity shown in Figure 9.5(d), a third CALMET run was made, in which the vertical extrapolation option was chosen in conjunction with the option of the O'Brien (1970) procedure, whereby vertical velocity is adjusted in such a way that it vanishes at the top of the model domain. Hence, in summary, three CALMET wind fields were generated based on these model options:

1. Vertical extrapolation (of surface wind observations)
2. No vertical extrapolation
3. Vertical extrapolation with O'Brien procedure

These three gridded CALMET wind fields were then used to run the CALPUFF, HPAC, and VLSTRACK dispersion models. Figures 9.7(a)-(d) show the scat-

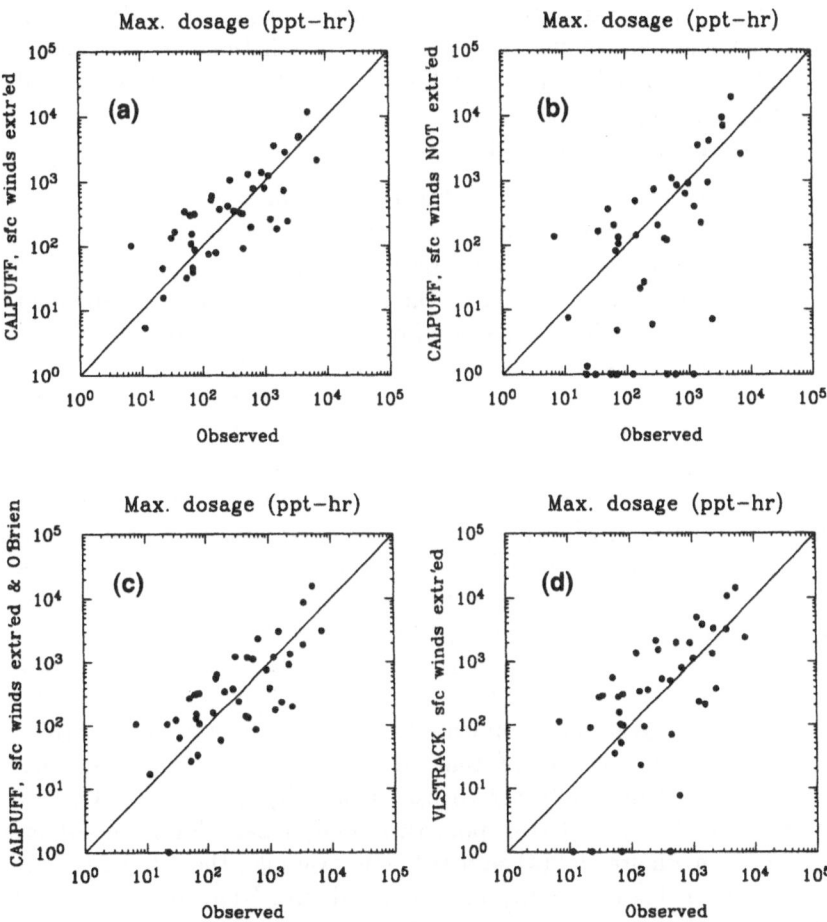

Fig. 9.7. Scatter plots of the maximum dosage (ppt-hr) at each sampling line for (a) CALPUFF coupled with the CALMET wind field in which the surface wind observations were vertically extrapolated, (b) CALPUFF coupled with the CALMET wind field without the extrapolation option, (c) CALPUFF coupled with the CALMET wind field with the options of vertical extrapolation and the O'Brien procedure, and (d) VLSTRACK coupled with the CALMET wind field with the extrapolation option.

ter plots of the predicted and observed maximum dosages along a sampling line for the above three CALMET wind fields coupled with CALPUFF, and the CALMET wind field for Option 1 coupled with VLSTRACK. Figure 9.7(a) is the same as Figure 9.3(a), and is for the baseline CALPUFF run. Figures 9.7(a) and (b) illustrate the previous argument that the more robust results associated with the baseline CALPUFF run were mainly due to the extrapolation of surface wind observations. Figure 9.7(c) shows that the invocation of the O'Brien procedure had only modest influence on the CALPUFF results (cf. Figure 9.7(a)). Figure 9.7(d) shows that the use of the CALMET wind field with the vertical extrapolation option seemed to worsen the VLSTRACK performance somewhat, in that the VLSTRACK predictions are now more scattered when compared to Figure 9.3(c). Thus, the CALMET wind field (i.e. Option 1) that improved the CALPUFF performance was not necessarily able to improve the VLSTRACK performance.

Figures 9.8(a), (c), and (d) show the HPAC predictions based on the above three CALMET wind fields. When compared to Figure 9.3(b), the baseline HPAC run, the new HPAC results based on the three CALMET wind fields are visibly much worse in terms of the number of outliers and the degree of scattering. This is similar to the case for VLSTRACK. This might be a result of the unrealistic vertical velocity field (see Figure 9.5(d)) predicted by CALMET after extrapolating surface wind observations. CALPUFF was not affected by this unrealistic vertical wind field, because the model's puff transport is only two-dimensional, i.e., the vertical velocity is not used by CALPUFF. However, both HPAC and VLSTRACK use the full three-dimensional velocity field to calculate puff transport. Therefore, strong vertical velocity might have an undesirable consequence of transporting puffs elsewhere. For example, the maximum HPAC prediction in Figure 9.3(b) is around 18000 ppt-hr (whereas the maximum observed is 7000 ppt-hr, see also Table 9.1). The maximum HPAC predictions in Figures 9.8(a), (c) and (d) are 2600, 34300, and 32500 ppt-hr, respectively. It is clear that strong vertical velocity led to much lower maximum concentration, as indicated in Figure 9.8(a). Figure 9.8(b) is the same as Figure 9.8(a), except that the CALMET vertical velocity fields at all levels were deliberately set to zero before being used by HPAC, hence effectively rendering puff transport two-dimensional. Noticeable differences exist between Figures 9.8(a) and (b). For example, the number of outliers is much less in Figure 9.8(b), and the maximum HPAC prediction is 22700 ppt-hr in Figure 9.8(b), almost an order of magnitude larger than that for Figure 9.8(a).

To further illustrate the sensitivity of the HPAC results to the CALMET vertical velocity field, Figure 9.9(a) shows the surface (1.5 m above the ground) dosage predicted by HPAC for Trial 3 (November 8, 1991) at 3.5 hours after the release, where surface wind observations were extrapolated vertically by CALMET. Figure 9.9(b) is the same as Figure 9.9(a), except that the vertical velocity fields at all levels were set to zero. HPAC predicted much more plume material at the ground when the vertical velocity fields were ignored.

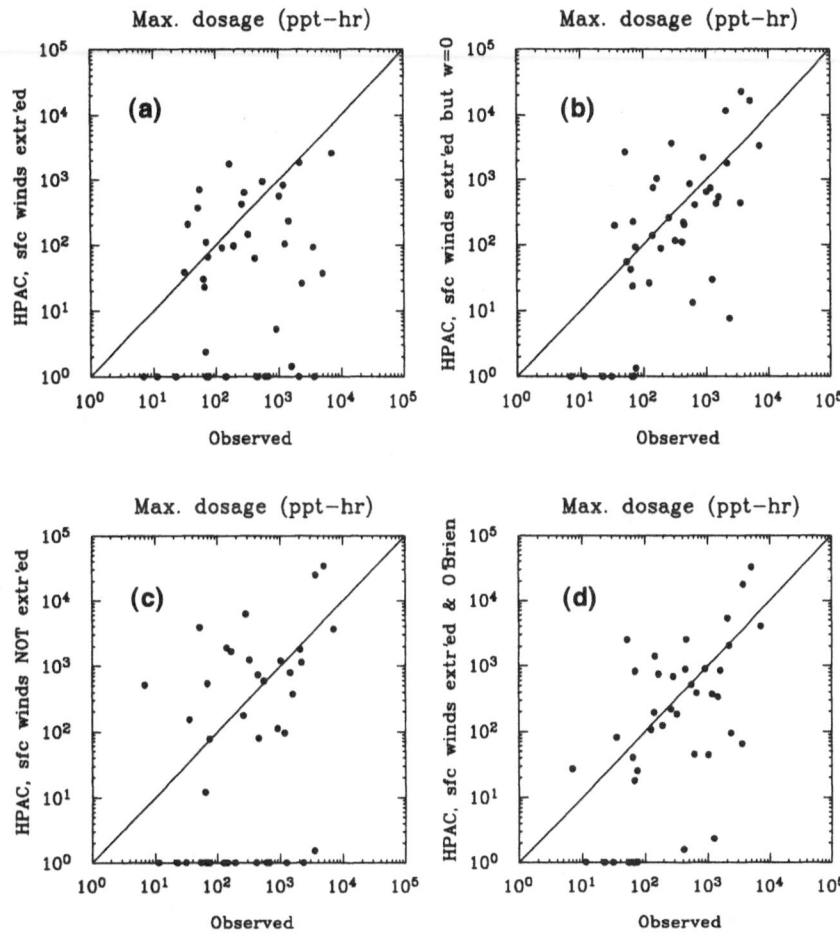

Fig. 9.8. Scatter plots of the maximum dosage (ppt-hr) at each sampling line for (a) HPAC coupled with the CALMET wind field in which the surface wind observations were vertically extrapolated, (b) same as (a) but the vertical velocity fields at all levels set to zero, (c) HPAC coupled with the CALMET wind field without the extrapolation option, and (d) HPAC coupled with the CALMET wind field with the options of vertical extrapolation and the O'Brien procedure.

Figure 9.10 shows the vertically integrated concentration predicted by HPAC for the same trial at one hour after the release for the same two variations of the CALMET wind fields. It can be seen that, unlike Figure 9.9, the contour areas for both cases are now much more comparable. This is because the vertically integrated results are less dependent on the height of the puff.

In conclusion, the second set of sensitivity studies further showed that the dispersion model results are extremely sensitive to the input meteorological

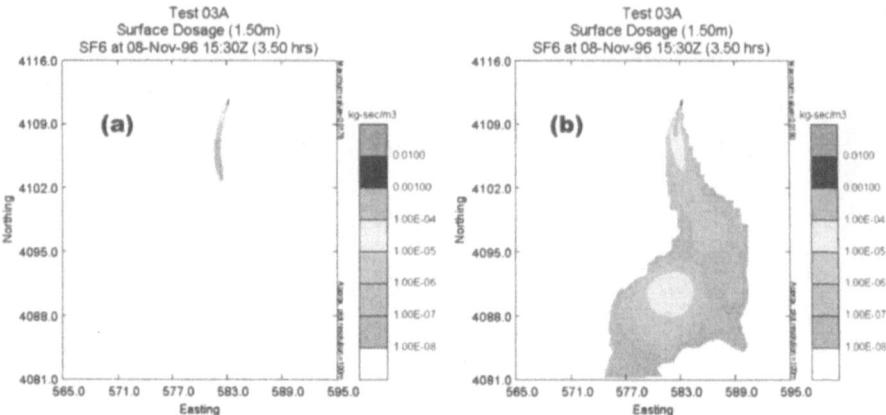

Fig. 9.9. Surface (1.5 m above ground dosage predicted by HPAC for DP26 Trial 3 (Nov 8, 1996) 3.5 hrs after release. (a) Based on the CALMET wind field in which surface wind observations were vertically extrapolated. (b) Same as (a) except that vertical velocity fields at all levels set to zero.

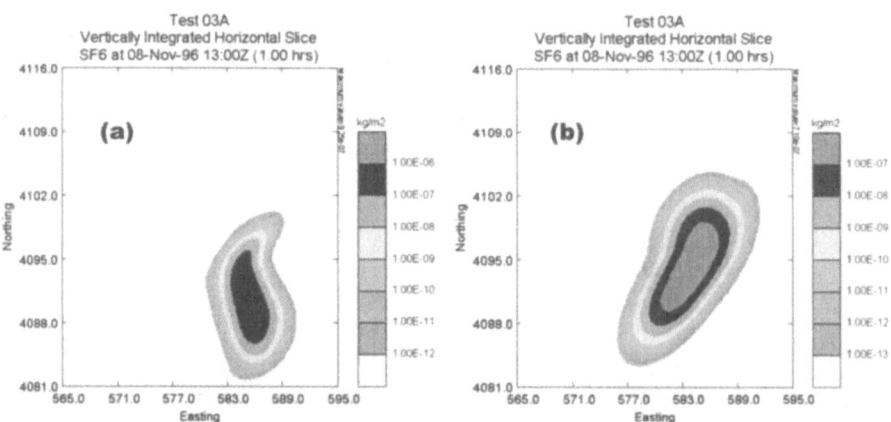

Fig. 9.10. Vertically integrated concentration predicted by HPAC for DP26 Trial 3 (Nov 8, 1996) 1.0 hrs after release. (a) Based on the CALMET wind field in which surface wind observations were vertically extrapolated. (b) Same as (a) except that vertical velocity fields at all levels set to zero.

fields, even when these fields were created by the same diagnostic model but with different model options. This dependence must be properly accounted for when interpreting the results. Finally, coupling the gridded meteorological fields prepared by one modeling system with another dispersion model might produce surprising results, due to the idiosyncrasies of each system.

9.3   Persian Gulf War Exposure Study

9.3.1   Background

The previous example illustrated a model evaluation and sensitivity study for a mesoscale domain (~30 km), where local meteorology was relatively well measured by eight surface stations and one upper-air station. SF_6 tracer gas was also carefully measured by a dense network of 90 whole-air samplers. Even with this rich set of onsite measurements, it has been demonstrated that the results are still quite sensitive to the gridded meteorological fields that dispersion models ultimately use. For 'real-world' scenarios such as accidental and tactical releases of toxic chemicals, the situation will be at the other end of the spectrum where onsite measurements of the source, meteorology, and exposure are not likely to be available. Moreover, the domain of interest might be larger that for DP26, e.g. in the order of hundreds of kilometers. Therefore, it would be of interest to study the issue of model uncertainty for these 'real-world' scenarios. In the following, the modeling of the 1991 Persian Gulf War will be used as an illustration.

It is now suspected that during and after the 1991 Persian Gulf War some chemical warfare agents stockpiled by Iraq might have been inadvertently released when the Coalition Forces destroyed various Iraqi ammunition storage sites through either ground detonations or aerial bombings (e.g. CIA/DoD 1997). Figure 9.11 shows a map of Iraq with surrounding countries. Because these potential releases were not identified until many years later, undoubtedly the source information was not well defined and concurrent local exposure measurements were not taken. Furthermore, routine meteorological measurements over Iraq were unavailable, because Iraq stopped reporting them to the World Meteorological Organization (WMO) after the 1981 Iran-Iraq war.

Dispersion models can estimate exposures of troops to these possible releases of chemical warfare agents. However, because Iraq as a whole was essentially devoid of any available meteorological observations, the prevailing meteorological conditions must be provided by *prognostic* mesoscale meteorological models. Pielke (1984) provides a good reference of mesoscale meteorological modeling. These models are based on fundamental conservation laws of mass, momentum, and energy. They are highly computationally intensive, and are capable of projecting meteorological conditions in both space and time with minimal observations. This is quite different from the *diagnostic* meteorological models considered in the case for DP26 (i.e. CALMET, SWIFT, and MC-SCIPUFF), where the models require sufficient local observations to perform spatial extrapolation and cannot forecast forward in time.

9.3.2   Overview of Prognostic Mesoscale Meteorological Models

Three state-of-the-art mesoscale meteorological models were used to reconstruct the weather conditions over Iraq during the Gulf War:

Fig. 9.11. Iraq with surrounding countries. Map courtesy the Perry-Castañeda Library Map Collection, University of Texas at Austin.

- COAMPS (Coupled Ocean-Atmospheric Mesoscale Prediction System; Westphal *et al.* 1999) developed by the Naval Research Laboratory (NRL)
- MM5 (Mesoscale Model, Version 5; Warner and Sheu 2000) developed by National Center for Atmospheric Research (NCAR) and Penn State University (PSU)
- OMEGA (Operational Multiscale Environmental Model with Grid Adaptivity; Bacon *et al.* 2000) developed by Science Applications International Corporation (SAIC)

The reader is referred to the respective reference for the details of each mesoscale meteorological model. In brief, COAMPS, MM5, and OMEGA are all three-dimensional mesoscale models that use similar parameterization schemes to simulate physical processes such as surface fluxes, moist convection, and shortwave and longwave radiation. COAMPS and MM5 are more similar in that they both use structured grids that can be multiply nested to focus on smaller areas of interest. The three nested grids (the outermost, second largest, and second smallest boxes) shown in Figure 9.12 for COAMPS and MM5 have grid spacings of 45, 15, and 5 km. OMEGA uses an unstructured grid that can dynamically adapt to the bottom topography. COAMPS and OMEGA apply the Level-2.5 planetary boundary layer scheme developed by Mellor and Yamada (1974). MM5 uses non-local PBL schemes that account for the influence of large, energetic eddies (Blackadar 1979, Hong and Pan 1996) Mesoscale models require observations to run. Observations include measurements from surface stations, rawinsonde observations, fixed and drifting buoys, civilian ship reports, pilot balloon observations, aircraft reports, temperature soundings from satellites, etc. Declassified observations were included when possible. Note that all in-situ observations were collected over areas surrounding Iraq. Other than a few isolated declassified observations collected by the U.S. Air Force Special Forces, the whole area of Iraq lacked any regularly-reported weather observations. This is the reason why prognostic models played a critical role in reconstructing the weather conditions over Iraq.

In addition to observations, mesoscale models also rely on large-scale (global) gridded fields to provide the initial and boundary conditions. These global fields are themselves products of global meteorological models. These global fields were used:

- The Navy Operational Global Atmospheric Prediction System (NOGAPS; Hogan and Rosmond 1991) for COAMPS, where the global fields are available at 1° grid interval and every six hours;

- The European Centre for Medium Range Forecast (ECMWF) global analysis archives for MM5, where the global fields are available at 2.5° grid interval and every 12 hours;
- The Global Optimal Interpolation (GOI) archives of the National Center for Atmospheric Research (NCAR) for OMEGA, where the global fields are available at 2.5° grid interval and every 12 hours.

Prognostic models can run in a pure 'forecast' mode or a 'four-dimensional data assimilation' (FDDA) mode (Daley 1991). In a pure forecast mode, the model numerically integrates from an initial time (t_0) only with the data available at t_0. The data available after t_0 are not used. In an FDDA mode, the model incorporates the data available after t_0 during the course of integration. From an operational point of view, daily weather forecasts are based on mesoscale models run in a forecast mode, because the data for, for example, tomorrow are not yet available. However, if prognostic models were to be applied to a historical period such as the 1991 Gulf War, then the data for

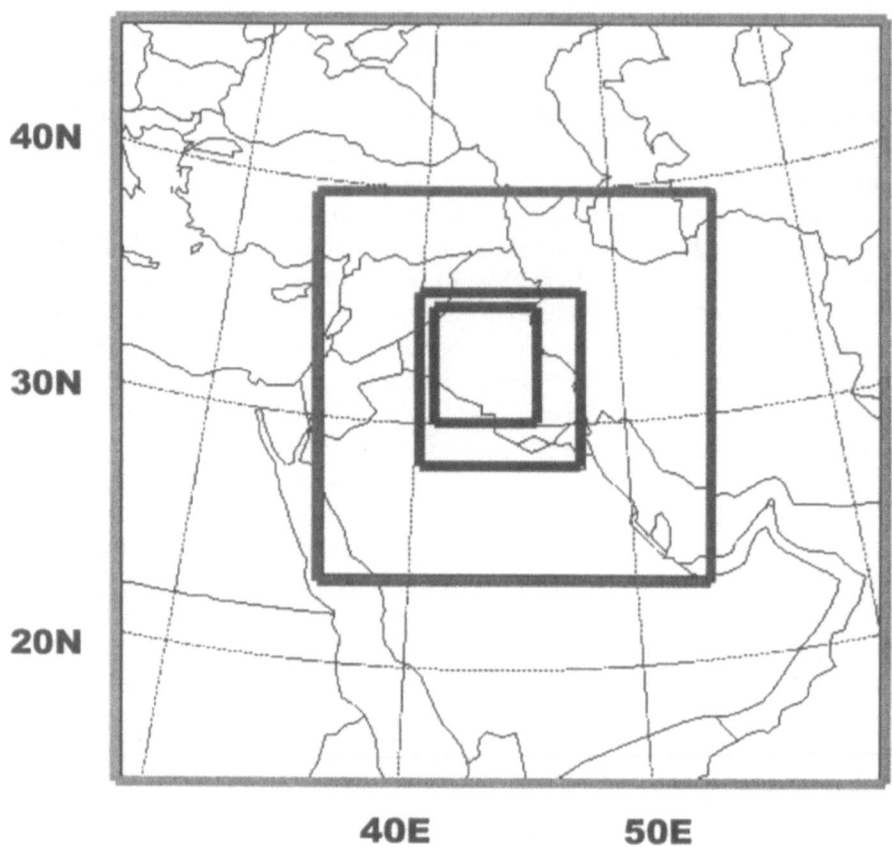

Fig. 9.12. Modelling domains for the three mesoscale meteorological models. Bounding frame, Grid 1 for COAMPS and MM5 (grid interval 45 km). Next largest (ie second) box, Grid 2 for COAMPS and MM5 (grid interval 15 km), and the OMEGA domain with varying grid interval. Third box (ie second smallest), Grid 3 for COAMPS and MM5 (grid interval 5 km). Evaluation of mesoscale models, ie predictions vs observations, was carried out over the second box. Inter-comparison of mesoscale models, ie predictions vs predictions, was carried out over the fourth (innermost) box that is primarily within Iraq.

the whole period would have already been available; as a result, the FDDA mode is possible and preferred. There are mainly two methods of performing FDDA. Intermittent FDDA performs data assimilation at distinct time intervals, e.g. every 12 hours, by means of objective analysis. Continuous FDDA assimilates observations continuously by modifying the basic governing equations solved by the models. COAMPS and OMEGA used intermittent FDDA, whereas MM5 used continuous FDDA.

A reanalysis, i.e. analysis of a historical period with the best available data, was performed by NRL (for COAMPS), NCAR (for MM5), and SAIC (for

OMEGA) for the two-month period between January 16 and March 15, 1991. The two-month reanalysis created a valuable and sizable database that contains high-resolution, three-dimensional meteorological conditions over the Persian Gulf region. These simulated data were then used as inputs to two dispersion models, VLSTRACK and HPAC.

9.3.3   Model Evaluation and Inter-Model Comparison

Since there were no meteorological measurements available over Iraq, the performance of the mesoscale meteorological models can only be evaluated with the available observations from countries surrounding Iraq.

In this study, all observations within the second (second largest) box in Figure 9.12 were considered for model performance evaluation. For areas inside Iraq, i.e. the fourth, innermost box in Figure 9.12, it is only possible to assess the differences among the three mesoscale meteorological models, with no reference to observations. Before more results are presented, it is helpful to first define the terms 'model evaluation' and 'inter-model comparison' as used here.

Term	Definition	Domain of interest
Model Evaluation	Compare model predictions with observations	*Second largest* box in Figure 9.12 (no observations available inside Iraq)
Inter-model Comparison	Compare predictions from one model with predictions from another model	*Innermost* box in Figure 9.12

The root mean square error ($RMSE$) was calculated to quantify model performance and inter-model difference.

$$RMSE = \sqrt{\frac{1}{N} \sum_{i=1}^{N} (M_i - O_i)^2} \qquad (9.7)$$

For the case of model evaluation, M_i and O_i denote the ith pair of model prediction and observation, and N is the total number of pairs. For the case of inter-model comparison, M_i and O_i denote the ith pair of model predictions. Surface measurements are usually taken at a height of 2 m above the ground. On the other hand, the lowest levels for COAMPS, MM5, and OMEGA are 10, 40, and 30 m, respectively, above the ground. When performing model evaluation and inter-model comparison, no attempts were made to extrapolate wind data to the same height. Therefore, it is likely that the MM5 surface wind speeds are consistently higher than the COAMPS surface wind speeds mainly due to different reporting heights.

Figure 9.13(a) shows the surface wind direction $RMSE$ of model predictions against observations for January 17 through February 25, 1991. The wind

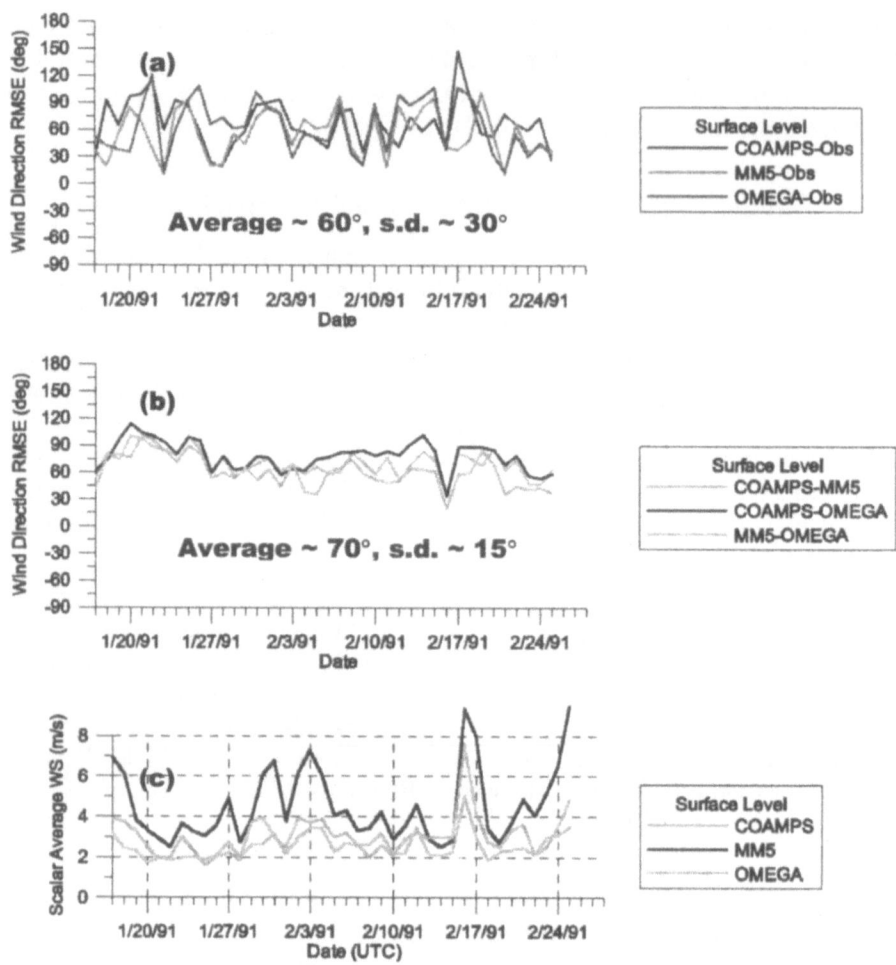

Fig. 9.13. Surface wind direction root mean square error ($RMSE$) for Jan 17 through Feb 25, 1991, for the three mesoscale meteorological models. (a) Model evaluation, ie predictions vs observations, over the second largest box area in Figure 9.12. (b) Inter-model comparison, ie predictions vs predictions, over the fourth (innermost) box area in Figure 9.12. (c) Scalar average surface wind speeds for the three models over the innermost area. The 'suface' levels for COAMPS, MM5 and OMEGA are 10, 40 and 30 m above ground, respectively.

direction uncertainty is in the range 30° to 90° This is consistent with Hanna and Yang (2000). The $RMSE$ time series for the three mesoscale models have an average of 60° and a standard deviation of 30° Figure 9.13(b) shows the inter-model variability in surface wind direction for areas inside Iraq (i.e. the red box in Figure 9.12). As a reference, Figure 9.13(c) shows the average surface wind speeds predicted by the three models for the same region. A general bias is clearly visible in Figure 9.13(c) due to height differences, with COAMPS showing the lowest wind speeds, and MM5 showing the highest wind speeds. Figures 9.13(b) and (c) are consistent with intuition that the wind direction variability is less for higher wind speeds, and more for lower wind speeds. It is also interesting to notice that while model evaluation shows the accuracy of surface wind direction predictions to fluctuate between 30 and 90° over areas outside Iraq (see Figure 9.13(a)), the inter-model variability for areas inside Iraq is relatively constant at 70° see Figure 9.13(b)). This means that when the wind fields simulated by the three mesoscale models are used to drive dispersion models, it is quite possible that the predicted plume trajectories will show large disparity. For example, one wind field might result in the toxic cloud moving to the west, whereas another wind field might result in the toxic cloud moving to the northeast. This uncertainty will have substantial impacts on the decision making process.

Figures 9.14(a) and (b) are the same as Figures 9.13(a) and (b), except that they are for surface wind speeds. The average surface wind speeds are again plotted, in Figure 9.14(c). The $RMSE$ for model evaluation, i.e. compared against observations, is in the order of 2 to 3 m/s. The $RMSE$ for inter-model comparison is roughly in the same magnitude, but with less day-to-day fluctuations. Part of the errors is due to different reporting heights.

As previously described, the total uncertainty in the model results can come from many sources, one of which is input uncertainty. Figure 9.13(b) shows that the typical uncertainty in predicted surface wind directions for areas inside Iraq is about 70° It would be interesting to attribute this wind direction uncertainty. The three mesoscale models essentially used the same observational data. However, there are many other areas where differences existed, such as modeling options, numerical schemes, data assimilation techniques, model grid structures, surface characteristics, and initial and boundary conditions.

In addition to measurements from randomly-spaced stations, mesoscale meteorological models also require gridded global meteorological model outputs as the initial and boundary conditions. The global fields used by COAMPS, MM5, and OMEGA were NOGAPS, ECMWF, and GOI, respectively, as previously stated. Figures 9.15(a) through (c) show the inter-model variability in the 1000-mb (near-surface) wind directions for the three global fields along the outermost boundary (the bounding frame) of Figure 9.12, where (a) is for wind direction $RMSE$, (b) is for wind speed $RMSE$, and (c) is for the average scalar wind speed. This boundary is the model domain for COAMPS and MM5. OMEGA, on the other hand, has a smaller domain as defined by the next largest box in Figure 9.12. Thus, strictly speaking, the results shown in Figure 9.15 do not represent a true comparison of the global wind fields used to provide the boundary conditions for the three mesoscale models. Nevertheless, Figure 9.15 still provides a qualitative description. The height of

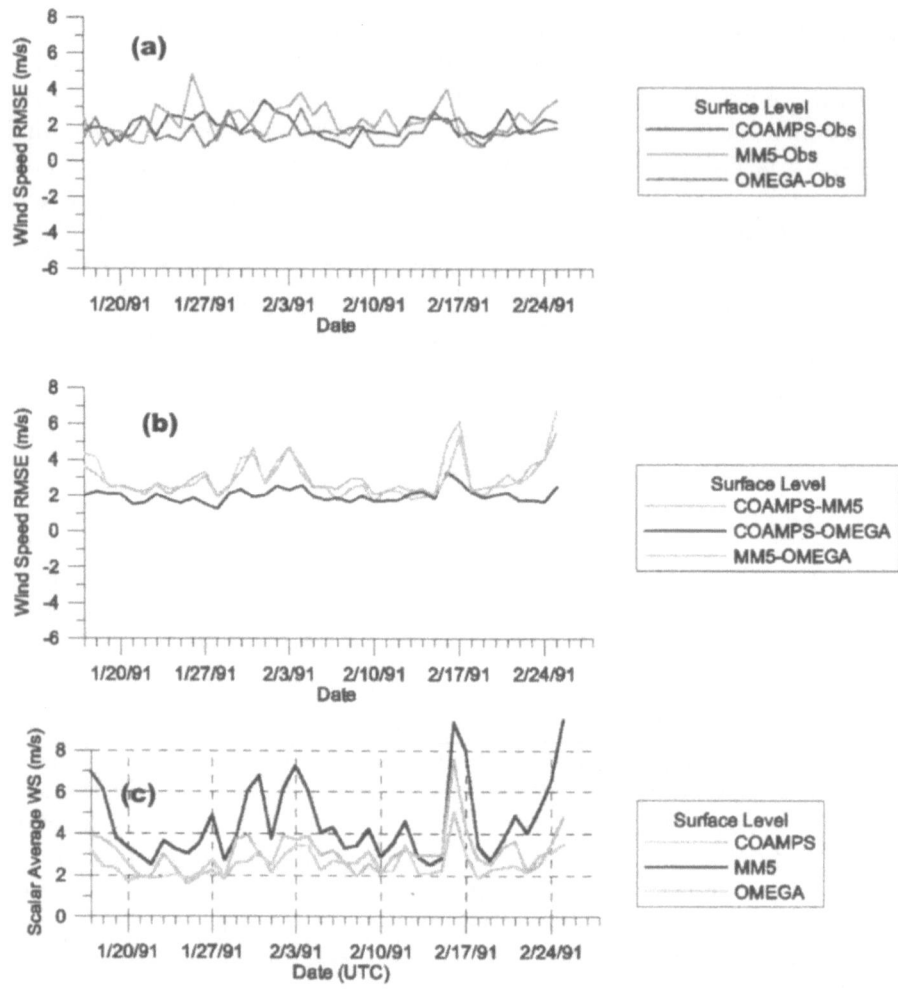

Fig. 9.14. Surface wind speed root mean square error (*RMSE*) for Jan 17 through Feb 25, 1991, for the three mesoscale meteorological models. (a) Model evaluation, ie predictions vs observations, over the second largest box area in Figure 9.12. (b) Inter-model comparison, ie predictions vs predictions, over the fourth (innermost) box area in Figure 9.12. (c) Scalar average surface wind speeds for the three models over the innermost area. The 'suface' levels for COAMPS, MM5 and OMEGA are 10, 40 and 30 m above ground, respectively.

1000 mb was selected because dispersion is likely to be controlled by near-surface wind fields. Figure 9.15(a) shows that the typical wind speed *RMSE* among the global fields along the boundary is about 2 to 3 m/s, and the wind direction *RMSE* is between 30° to 60° Therefore, the large (∼70° surface wind direction *RMSE* found earlier (see Figure 9.13(b)) for areas inside Iraq

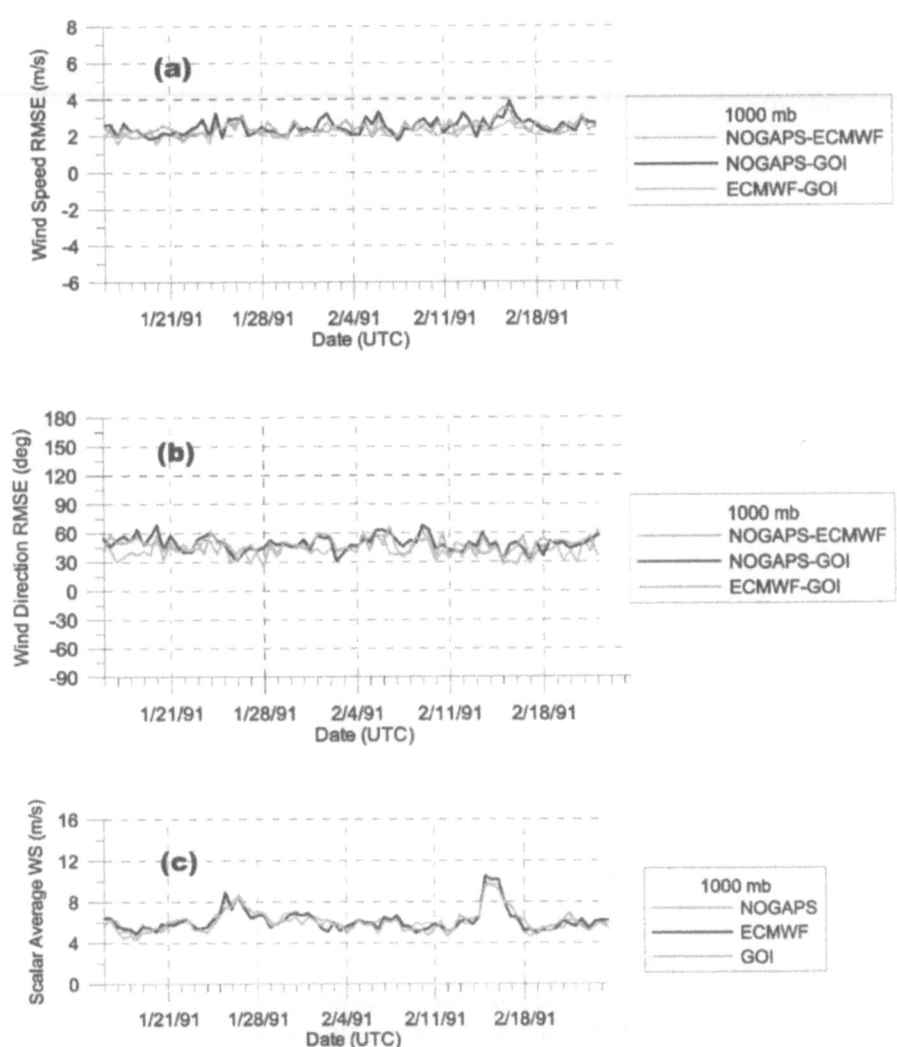

Fig. 9.15. Inter-model comparison, i.e. predictions vs predictions, along the perimeter of the outermost (i.e. bounding) box in Figure 9.12 for the three *global*-scale models. mesoscale meteorological models. (a) 1000-mb (near-surface) wind speed root mean square (*RMSE*). (b) 1000-mb wind direction *RMSE*. (c) 1000-mb scalar average wind speeds.

can be partly explained by the differences that already existed in the global fields.

To demonstrate the consequence of using the three mesoscale wind fields to drive dispersion models when no onsite data were available, assume two hypothetical releases that took place at 0200 UTC, January 19, 1991; and at

0100 UTC, January 31, 1991. The release location was assumed at 33.3°N, 42.7°E. For simplicity, assume an instantaneous release of 180 kg of chemical warfare agents. Consider a dosage of 0.0432 mg-min/m³ accumulated over 24 hours to define the plume boundary. Figure 9.16 shows the 24-hr dosage

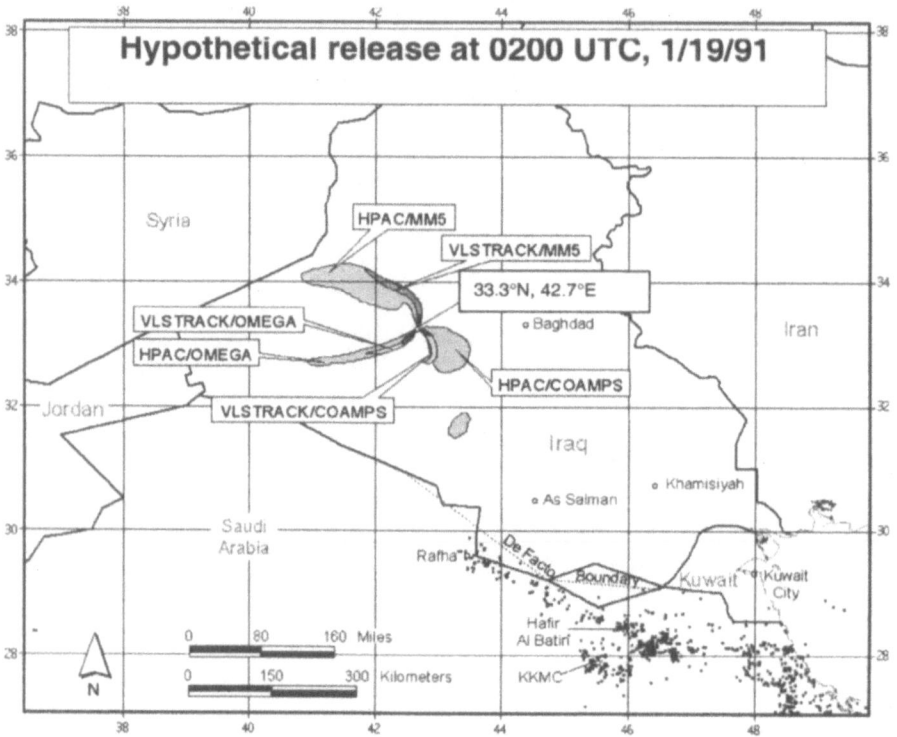

Fig. 9.16. Dosage contours (0.0432 mg-min/m³ accumulated over 24 hrs) predicted by HPAC and VLSTRACK, coupled with COAMPS, MM5 and OMEGA wind fields for a hypothetical release occurring at 0200 UTC, Jan 19, 1991. Dots represent US troop locations.

contours as predicted by HPAC and VLSTRACK driven by the COAMPS, MM5, and VLSTRACK wind fields for the January 19 release. Two dispersion models and three meteorological models yielded six predicted dispersion patterns. Disparity in plume footprints is great; particularly, the footprints based on the MM5 and COAMPS wind fields seem to be opposite to each other. As discouraging and unsatisfactory as it might seem, the difference *is* consistent with the uncertainty in predicted surface wind directions. Figure 9.17 shows the results for the January 31 release, where the agreement is much better. Warner *et al.* (2000) also reported similar findings.

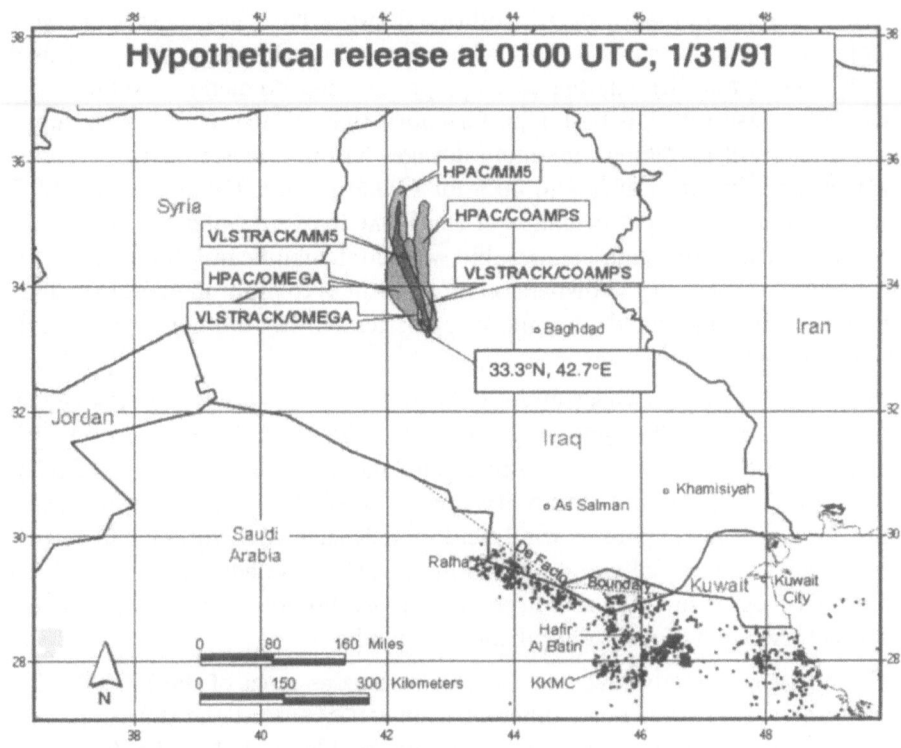

Fig. 9.17. Dosage contours (0.0432 mg-min/m^3 accumulated over 24 hrs) predicted by HPAC and VLSTRACK, coupled with COAMPS, MM5 and OMEGA wind fields for a hypothetical release occurring at 0100 UTC, Jan 31, 1991. Dots represent US troop locations.

9.4   Conclusions

It is important to interpret dispersion modeling results within the context of model sensitivity and uncertainty. Uncertainty in dispersion modeling can come from random turbulence, data errors, and model errors. Some errors are reducible, and some are not. Errors from different sources will accumulate. Two case studies were presented in this paper, the Dipole Pride 26 (DP26) field experiment and the Gulf War exposure assessment, to study model sensitivity and uncertainty. The main focus was on meteorological input data, and the results were primarily presented in a qualitative fashion.

The first case study involved a mesoscale (~30 km) field experiment where the source, meteorological, and concentration data were well measured by a dense network. The observed surface wind fields sometimes showed strong spatial variability due to surrounding mountains. The CALPUFF, HPAC, and VLSTRACK dispersion models were first run in a 'typical' fashion where each model was allowed to prepare the necessary gridded wind fields from the same raw meteorological data. Additional two HPAC sensitivity runs were made with different choices of (i) which HPAC built-in diagnostic wind field model to use, and (ii) whether to use upper air measurements. CALPUFF and the HPAC base run had better performance than VLSTRACK. For example, the fractions of predicted maximum dosages within a factor of two of observations were between 40% and 50% for CALPUFF and HPAC, but was 20% for VLSTRACK. The performance of the HPAC base run and the two HPAC sensitivity runs, although comparable, was often significantly different. This suggests the sensitivity of dispersion model results to input meteorology. Furthermore, running models in a more sophisticated way did not always result in better performance.

A second set of DP26 sensitivity studies was performed where all dispersion models were driven by common gridded CALMET wind fields, prepared with various model options. The results again suggested great model sensitivity, partly due to the spatial variability present in observed surface wind fields, and the fact that diagnostic wind field models such as CALMET had to reconcile this large spatial variability. Moreover, the gridded wind field that led to more robust results for one dispersion model did not necessarily guarantee satisfactory results for another dispersion model.

The second case study involved the exposure assessment of the 1991 Persian Gulf War. This case is quite different from DP26 in that the source, meteorological, and concentration data were unavailable; and that the domain of interest was in the order of hundreds of kilometers. Prevailing meteorological conditions over Iraq had to be reanalyzed with state-of-the-art mesoscale prognostic meteorological models, based on weather observations from countries surrounding Iraq and outputs from global-scale meteorological models. The performance of these mesoscale models can only be evaluated with data from areas outside Iraq, i.e. apart from the main domain of interest. It was found that predicted surface wind directions had errors, when compared to observations, between 30° to 90°. Over Iraq, i.e., inside the main domain of interest, only the uncertainty among the mesoscale models can be assessed. The inter-model variability in predicted surface wind directions was fairly constant at 70° regardless of the prevailing large-scale flow conditions. This large uncertainty in simulated surface wind fields has major impacts on predicted dispersion patterns, and is unlikely to be further reduced due to (i) lack of onsite meteorological data and (ii) random turbulence in the atmosphere. Two hypothetical releases were considered. While the three mesoscale wind fields produced similar dosage contours for one release, the dosage contours for the other release showed large disparity. It was further demonstrated that part of the uncertainty in the mesoscale wind fields already existed in the global fields that were used as the initial and boundary conditions for the mesoscale models.

In conclusion, the above two case studies, although quite different in their characteristics, are all typical applications of dispersion modeling. For the well-instrumented Dipole Pride 26 field experiments, it was found that model performance was very sensitive to how meteorological data were specified. More sophisticated ways of running a model did not guarantee better model performance. A robust wind field for one dispersion model might not be satisfactory for another dispersion model. For the Persian Gulf War exposure study where there were no onsite meteorological data, the uncertainty in predicted surface wind directions was as large as 70°. This sometimes led to large disparity in the locations of predicted hazard areas. As a consequence, it is imperative that any dispersion model results must be interpreted with a proper expectation in the light of model sensitivity and uncertainty.

Acknowledgments

This study was partly supported by the Defense Threat Reduction Agency (DTRA) and the Office of the Special Assistant to the Under Secretary of Defense (Personnel and Readiness) for Gulf War Illnesses, Medical Readiness, and Military Deployments. Mesoscale meteorological modeling for the Persian Gulf War study was performed by the Naval Research Laboratory (NRL) for COAMPS, National Center for Atmospheric Research (NCAR) for MM5, and Science Applications International Corporation (SAIC) for OMEGA. Dispersion modeling for the Persian Gulf War study was performed by DTRA for HPAC; and the Naval Surface Warfare Center (NSWC), the Dahlgren Division, for VLSTRACK. The U.S. Army Center for Health Promotion and Preventive Medicine (USACHPPM) prepared the original versions of Figures 9.16 and 9.17. The views expressed in this article are those of the author and do not reflect the official policy or position of the Under Secretary of Defense (Personnel and Readiness) for Gulf War Illnesses, Medical Readiness, and Military Deployments; or the Department of Defense.

References

Anthes, R.A., Kuo, Y.-H., Hsie, E.-Y., Low-Nam, S. & Bettge, T.W. (1989). Estimation of skill and uncertainty in regional numerical models. *Quarterly J. Royal Meteorol. Soc.*, **115**, 763-806.

Arya, S.P. (1999). *Air Pollution Meteorology and Dispersion*. Oxford University Press, 310pp.

Bacon D., Ahmad, N., Boybeyi, Z., Dunn, T., Hall, M.,Lee, P., Sarma, R.A., Turner, M., Waight, K., Young, S. & Zack, J. (2000). A dynamically adapting weather and dispersion model: the Operational Multiscale Environment Model with Grid Adaptivity (OMEGA). *Monthly Weather Review*, **128**, 2044-2076.

Bauer, T.J. & Gibbs, R.L. (1998). *Software User's Manual for the Chemical/Biological Agent Vapor, Liquid, and Solid Tracking (VLSTRACK) Computer Model, Version 3.0*. Systems Research and Technology Department,

Dahlgren Division, Naval Surface Warfare Center, Dahlgren, VA 22448-5100.

Bergin, M.S., Noblet, G.S., Petrini, K., Dhieux, J.R., Milford, J.B. & Harley, R.A. (1999). Formal uncertainty analysis of a Lagrangian photochemical air pollution model. *Environ. Sci. Techonol.*, **33**, 1116-1126.

Biltoft, C.A. (1998). Dipole Pride 26: *Phase II of Defense Special Weapons Agency Transport and Dispersion Model Validation.* Prepared for Defense Special Weapons Agency, 6801 Telegraph Road, VA 22310, by Meteorology & Obscurants Divisions, West Desert Test Center, U.S. Army Dugway Proving Ground, Dugway, UT 84022.

Blackadar, A.K. (1979). High resolution models of the planetary boundary layer. *Advances in Environmental Science and Engineering,* eds. Pfafflin, J. & Ziegler, E., Gordon and Breach, 50-85.

Carslaw, N., Jacobs, P.J. & Pilling, M.J. (1999). Modeling OH, HO_2, and RO_2 radicals in the marine boundary layer, Part 2, mechanisms reduction and uncertainty analysis. *J. Geophys. Res.*, **104**, No. D23, 30257-30273.

Chang, J.C., Franzese, P. & Hanna, S.R. (1999). *Evaluation of* CALPUFF, HPAC, *and* VLSTRACK *with the Dipole Pride 26 Field Data.* School of Computational Sciences, George Mason University, MSN 5C3, Fairfax, VA 22032.

Chatwin, P.C. & Sullivan, P.J. (1979). The relative diffusion of a cloud of passive contaminant in incompressible turbulent flow. *J. Fluid. Mech.*, **91**, 337-355.

CIA/DoD (1997). *Modeling the Chemical Warfare Agent Release at the Khamisiyah Pit.* Central Intelligence Agency and Department of Defense. Available from the Office of the Special Assistance for Gulf War Illnesses, 5113 Leesburg Pike, Suite 901, Falls Church, VA 22041. 30 pp.

Csanady, G.T. (1967). Concentration fluctuations in turbulent diffusion. *J. Atmos. Sci.*, **24**, 21-28.

Daley, R., 1991: *Atmospheric Data Analysis*, Cambridge University Press, 457pp.

Debberdt, W.F. & Miller, E. (2000). Uncertainty, ensembles, and air quality dispersion modeling: applications and challenges. *Atmos. Environ.*, **34**, 4667-4673.

DTRA (1999). HPAC *Hazard Prediction and Assessment Capability, Version 3.2*, Defense Threat Reduction Agency, 6801 Telegraph Road, VA 22310.

Efron, B. (1987). Better bootstrap confidence intervals. *J. Amer. Statist. Assoc.*, **82**, 171-185.

Fish, D.J. (2000). The automatic generation of reduced mechanisms for tropospheric chemistry modeling. *Atmos. Environ.*, **34**, 1563-1574.

Fox, D.G. (1984). Uncertainty in air quality modeling. *Bulletin of Amer. Meteorol. Soc.*, **65**, 27-36.

Gifford, F.A. (1959). Statistical properties of a fluctuating plume dispersal model. *Adv. Geophys.*, **6**, 117-137.

Hanna, S.R. & Yang, R. (2000). Evaluations of mesoscale model predictions of near-surface winds, temperature gradients, and mixing depths. To appear in *J. Appl. Meteorol.*

Hanna, S. R., Chang, J. C. & Fernau, M. E. (1998). Monte Carlo estimates of uncertainties in predictions by a photochemical grid model (UAM-IV) due to uncertainties in input variables. *Atmos. Environ.*, **32**, 3619-3628.

Hanna, S.R., Lu, Z., Frey, H.C., Wheeler, N., Vukovich, J., Arumachalam, S. & Fernau, M. (2000). Uncertainties in predicted ozone concentration due to input uncertainties for the UAM-V photochemical grid model applied to the July 1995 OTAG domain. To appear in *Atmos. Environ.*

Hanna, S.R., Chang, J.C. & Strimaitis, D.G. (1993). Hazardous gas model evaluation with field observations. *Atmos. Environ.*, **27A**, 2265-2285.

Hanna, S.R., Strimaitis, D.G. & Chang, J.C. (1991). *Hazard Response Modeling Uncertainty (A Quantitative Method), Volume I: User's Guide for Software for Evaluating Hazardous Gas Dispersion Models; Volume II: Evaluation of Commonly-Used Hazardous Gas Dispersion Models; Volume III: Components of Uncertainty in Hazardous Gas Dispersion Models.* Report no. A119/A120, prepared by Earth Tech, Inc., 196 Baker Avenue, Concord, MA 01742, for Engineering and Services Laboratory, Air Force Engineering and Services Center, Tyndall Air Force Base, FL 32403; and for American Petroleum Institute, 1220 L Street, N.W., Washington, D.C., 20005.

Hanna, S.R., Briggs, G.A. & Hosker, R.P. (1982). *Handbook on Atmospheric Diffusion.* U.S. Department of Energy, Technical Information Center, Oak Ridge, Tennessee.

Hogan, T. & Rosmond, T. (1991). The description of the Navy Operational Global Atmospheric Prediction System's spectral forecast model. *Monthly Weather Review*, **119**, 1786-1815.

Hong, S.-Y. & Pan, H.-L. (1996). Nonlocal boundary layer vertical diffusion in a medium-range forecast model. *Monthly Weather Review*, **124**, 2322-2339.

Kaplan, H. & Dinar, N. (1993). A three-dimensional model for calculating the concentration distribution in inhomogeneous turbulence. *Bound. Layer Meteorol.*, **62**, 217-245.

Mellor G. L. & Yamada, T. (1974). A hierarchy of turbulence closure models for planetary boundary layers. *J. Atmos. Sci.*, **31**, 1791-1806.

O'Brien, J.J. (1970). A note on the vertical structure of the eddy exchange coefficient in the planetary boundary layer. *J. Atmos. Sci.*, **27**, 1213-1215.

Pielke, R.A. (1984) *Mesoscale Meteorological Modeling.* Academic Press, 612pp.

Sawford, B.L. (1983). The effect of Gaussian particle-pair distribution functions in the statistical theory of concentration fluctuations in homogeneous turbulence. *Quarterly J. Royal Meteorol. Soc.*, **109**, 339-354.

Scire, J.S., Strimaitis, D.G. & Yamartino, R.J. (2000a). *A User's Guide for the* CALPUFF *Dispersion Model (Version 5.0)*. Earth Tech, Inc., 196 Baker Avenue, Concord, MA 01742.

Scire, J.S., Robe, F.R., Fernau, M.E. & Yamartino, R.J. (2000b): *A User's Guide for the* CALMET *Meteorological Model (Version 5.0)*. Earth Tech, Inc., 196 Baker Avenue, Concord, MA 01742.

Stensrud, D.J., Bao, J.-W. & Warner, T.T. (2000). Using initial condition and model physics perturbations in short-range ensemble simulations of mesoscale convective systems. *Monthly Weather Review*, **128**, 2077-2107.

Straume, A.G., Koffi, E.N. & Nodop, K. (1998). Dispersion modeling using ensemble forecasts compared to ETEX measurements. *J. Appl. Meteorol.*, **37**, 1444-1456.

Sykes, R.I., Lewellen, W.S. & Parker, S.F. (1984). A turbulent-transport model for concentration fluctuations and fluxes. *J. Fluid Mech.*, **139**, 193-218.

Sykes, R.I. & Henn, D.S. (1992). Large-eddy simulation of concentration fluctuations in a dispersion plume. *Atmos. Environ.*, **26A**, 3127-3142.

Warner, T.T., Sheu, R.-S., Bowers, J.F., Sykes, R.I. & Dodd, G.C. (2000). Ensemble atmospheric dispersion simulations employing a coupled atmospheric dynamic model and a dispersion model. *Proceedings of Fourth George Mason University Transport and Dispersion Modeling Workshop*, George Mason University, Fairfax, VA, 22030.

Warner, T. T. & Sheu, R.-S. (2000). Multiscale local forcing of the Arabian Desert daytime boundary layer, and implications for the dispersion of surface-released contaminants. *J. Appl. Meteorol.*, **39**, 686-707.

Watson, T.B., Keislar, R.E., Reese, B. , George, D.H. & Biltoft, C.A. (1998). *The Defense Special Weapons Agency Dipole Pride 26 Field Experiment*. NOAA Technical Memorandum ERL ARL-255, Air Resources Laboratory, Silver Spring, Maryland.

Westphal, D.L., Holt, T.R., Chang, S.W.,Baker, N.L., Hogan, T.F., Brody, L.R., Godfrey, R.A., Goerss, J.S., Cummings, J.A., Laws, D.J. & Hines, C.W. (1999). Meteorological reanalysis for the study of Gulf War illnesses: Khamisiyah case study. *Weather and Forecasting*, **14**, 215-241.

Willis, G.E. & Deardorff, J.W. (1976). A laboratory model of diffusion into the convective boundary layer. *Quarterly J. Royal Meteorol. Soc.*, **102**, 427-445.

Willis, G.E. & Deardorff, J.W. (1978). A laboratory study of dispersion from an elevated source within a modeled convective planetary boundary layer. *Atmos. Environ.*, **12**, 1305-1311.

Wilson, D.J. (1994). *Concentration Fluctuations in Toxic and Flammable Vapor Cloud Dispersion: Recommendations for an Operational Hazard Assessment Model.* CCPS Project No. 73, prepared for the Vapor Cloud Sub-committee, AIChE, Center for Chemical Process Safety, 345 East 47 Street, New York, NY 10017.

## Part IV

Risk and Uncertainty

10 Statistics and the Environmental Sciences: Approaches to Model Combination

Gudmund Høst

Norwegian Computing Center,
P.O. Box 114, Blindern
N-0314 Oslo, Norway

Abstract. When making statistical assessments in the environmental sciences, we may have multiple sources of information available for inference. For example, we may have access to both observations and a mechanistic model for describing the process of interest. The modeling view taken from the physical sciences is traditionally very different from the statistical view. Bayesian filtering methods and state-space models are capable of making use of dynamics and measurements within a coherent mathematical framework. This methodology has become commonplace within the atmospheric and oceanographic sciences, although practical application poses large computational and numerical problems. In contrast, we may in some situations not have direct access to the physical model and the fitting of its parameters, but we would still like to take advantage of this information in making the probability assessment. In this paper, a statistical framework for combining models from physics and statistics in the space–time domain is introduced. Bayesian and non-Bayesian approaches to such combination are described and discussed. Illustrations are given in terms of applications within long range air pollution and flood forecasting.

10.1 Introduction

The main purpose of this paper is to introduce a unified statistical framework for combination of statistical and physical models and then use this framework to compare and discuss various approaches to solving this problem. In the following, a statistical model is defined by the specification of the likelihood function, while a physical model typically is described by conservation equations derived from 'first principles'. A statistical model typically describes the observed data, while a physical model describes some underlying process that generates the data. The term 'physical' will be used here in a wide and generic sense, also incorporating models within meteorology, oceanography or, say, chemistry.

The need for such model combination arises naturally in almost any context where measurements are to be combined with process-understanding. A typical example is weather forecasting, where output from a meteorological computerized model are to be combined with a monitoring network and remotely sensed observations to give the best forecast of future weather. One approach is to build a more general model that combines the likelihood from the data with the information from the meteorological model. Another example is flood forecasting, where a hydrological rainfall–runoff model may be adjusted to river flow observations to give improved flood forecasts.

Within applied statistics, the need to combine models arises in several different contexts. One such context is when various independent data sources on a common underlying parameter or process are available. In this case, the major challenge is for each data source to construct a statistical model in terms of a likelihood function that describes the observed variability of the data and involves a parametrization of the underlying process. In contrast, the combination of these statistical models is merely to form the product of these likelihood components. Another variant is meta-analysis, widely used within medicine and health studies (Olkin 1996). In meta-analysis, previously performed statistical experiments are combined to assess the effectiveness of alternative treatments. There are both frequentist and Bayesian approaches to meta-analysis. Cox & Piegorsch (1996) and Piegorsch & Cox (1996) review environmental applications of data combination.

The Bayesian approach is often a convenient and powerful framework for model combination. Here, the physical model state is regarded as prior information and a likelihood model is built for the observations. Bayesian computational tools may provide an estimate of the full posterior distribution or summary statistics of this distribution, such as the mode and curvature. In spatio-temporal problems, it is often necessary to do the calculations iteratively, which can be computationally very demanding. In the atmospheric and oceanographic sciences, this is often referred to as data assimilation (Miller *et al.* 1999, Evensen 1994), while filtering may be a more familiar term within other fields of research.

The structure of this paper is as follows. Some unifying notation and a statistical framework for this study is given in Section 10.2. Various approaches are illustrated on real data examples in Section 10.3 and discussed in Section 10.4.

10.2   Statistical Framework

Denote by $X(t)$ a hidden state of interest at time t in m–dimensional state space and denote by $y(t)$ an n–vector of observations. In practice, X may consist of one or more physical processes on a discrete spatial grid and y may be measurements from a related observable process. The joint probability density of X and y is $p(x, y)$. Furthermore, $p(y|x)$ is the likelihood function and $p(x|y)$ is the posterior density of X given y. Usually, X is the object under study that we want to make inference about. It may be described by a physical model or by some structured stochastic model. Standard probability calculus gives Bayes' rule, in the form of

$$p(x|y) = \frac{p(y|x)p(x)}{p(y)}. \tag{10.1}$$

Here, $p(x)$ is the prior density of x and $p(y)$ is a constant. Usually, $p(x)$ is not available to the statistician and should be specified from expert knowledge. This is the general setup for the classical Bayesian analysis, where the purpose is to compute the posterior density of X given the prior and the likelihood. The interpretation is that one will use the observations y to improve upon the prior knowledge of the process. To undertake such an analysis, access to the model X and its parameters is needed, so that the model may be altered according to what is learned from y. The problem is more complicated for space-time data, because the posterior has to be updated recursively in time. This is the general filtering problem (Doucet *et al.* 2000), and it is within the atmospheric sciences often referred to as data assimilation (Miller *et al.* 1999, Evensen 1994).

Another view is to regard X as a deterministic function, or equivalently, take $p(x)$ to be a delta function. In practical problem solving, the statistician may be given the output from a physical model X, and may not have the opportunity to alter this model. The object of the study is often to improve on the inference of y given $X = x$, therefore we are interested in $p(y|x)$. Other applications include forecast verification; see for example Murphy (1995) or the references therein. This approach is one of *conditional analysis*, where X is seen as fixed in the same manner as an external covariate. In the following, we will see that each of these approaches lead to diverse and challenging statistical analyses.

Conditional Analysis

To be specific, we give an account of conditional analysis in terms of Gaussian processes. A continuous version of the physical model can be defined through

$$\widetilde{X}_h(s, t) = \mathbf{1}' \, \mathbf{K}_h(s) \, \mathbf{X}(t). \tag{10.2}$$

Here, $\mathbf{K}_h(s)$ is an $m \times m$ diagonal matrix of weights with ith element

$$[\mathbf{K}_h(w)]_i = h^{-2} K[(s_i - w)/h],$$

and $K(\cdot)$ is an asymmetric density function supported on $(-1, 1) \times (-1, 1)$. Furthermore, h is the grid size used in the physical model. Usually, the process x is smooth over the spatial grid. Therefore, the construction of $\widetilde{X}_h(s, t)$ is not sensitive to the choice of $K(\cdot)$. In practice, X may consist of several interrelated processes discretized on the grid, but we will suppress this in our notation. In space-time, we may write

$$Y(s, t)|x = \widetilde{X}_h(s, t) + \varepsilon(s, t),$$

$$\mathrm{E}\,(Y(s, t)|\mathbf{X} = x) = \widetilde{X}_h(s, t), \tag{10.3}$$

$$\mathrm{Cov}\,(Y(s', t'), Y(s'', t'')|\mathbf{X} = x) = c\,(\|(s', t') - (s'', t'')\|\,; \mathbf{X} = \boldsymbol{\theta})\,.$$

Here $\varepsilon(s, t)$ is a spatio-temporal Gaussian random field with mean zero and covariance function $c(\cdot; \boldsymbol{\theta})$. In practice, the data may have been transformed to an approximate linear scale by some nonlinear transformation, such as the logarithm. The interpretation of (10.4) is that the physical model provides the trend or systematic component of the observable process, while the residual term accounts for the deviation between the trend and the true process. The residual field ε may account for variability of Y at spatial scales smaller than the grid resolution and temporal scales smaller than the time resolution of the physical model. The residual field will also incorporate model errors due to inadequate description of the observable process. Effects not explicitly described by the physical model may be mitigated into the residual term. Thus, the statistical model (10.4) may be used as a starting point for criticism of the physical model. For example, a 'good' physical model should explain most of the structure at the spatio-temporal scales that are resolved by its discretization. For such physical models, we would expect residual variability to be negligible at spatio-temporal lags greater than the resolution of the physical model. Of course, the statistical model (10.4) may easily be extended to incorporate terms accounting for systematic (non-random) model errors, which would be useful in verification studies of the physical model (Murphy 1995).

The study of the residual space-time field is the main focus of the statistical modeling exercise in a conditional analysis. Do the observations and the physical model describe the same process, and to what extent should these be similar? Fluid dynamics models are usually built by averaging the 'true' underlying random space-time field over the grid resolution, giving differential equations for the first order statistical moments of these fields. Due to non-linearity of the governing physical equations, these equations for the average field involve second order statistical moments. This involves quantification of space-time covariances and cross-covariances at the sub-grid scale. Modeling these second order moments from physics involves third order statistical moments, and so on. This classical *closure problem* (Tritton 1977) is a critical point in model implementation. Most geophysical processes are non-linear, and this implies that small-scale variability may transfer into the larger scales as time evolves.

An important advantage of using a physical model, as opposed to doing purely empirical modeling, is the accessibility of a physically-based explanatory tool for the data. In the example of European transboundary air pollution, this is of high importance, as it relates emissions to depositions and effects. The role of the statistical model is here to give uncertainty estimates and to quantify local variability. In the completely different example of flood forecasting, two deterministic models are run sequentially to give flood forecasts from weather forecasts. Here, the statistical supermodel is built to provide probability statements for the flood forecasts, as well as attributing uncertainty to each of the deterministic models.

Bayesian Analysis

For a Bayesian analysis in the Gaussian case, we may use the same model for $Y|x$ as specified in (10.4). In addition, we need a prior on X. In space-time, this prior describes the dynamic physical model in stochastic form. Often, we may need to solve a set of coupled dynamic equations for the physical processes involved. In this case, X would include the state vector of all these physical processes. We may describe the state vector by a stochastic differential equation

$$dX(t) = b(X(t))dt + \sigma(X(t))dW(t). \tag{10.4}$$

Here, $b(\cdot)$ is a drift function, $\sigma(\cdot)$ is the diffusion term (matrix) and $dW(t)$ is a Brownian motion. The diffusion process (10.4) has a corresponding transition equation, also known as the Kolmogorov forward equation or the Fokker-Planck equation,

$$\frac{\partial p(x,t)}{\partial t} = L^* p(x,t). \tag{10.5}$$

Here, $p(x,t)$ is the probability density of the m-dimensional state vector and the infinitesimal generator L^* is the adjoint of the operator

$$L(\cdot) = \sum_{i=1}^{m} b_i(\cdot) \frac{\partial}{\partial x_i}(\cdot) + \sum_{i,j=1}^{m} \frac{\partial^2}{\partial x_i \partial x_j} (\sigma \sigma^T)_{ij}(\cdot). \tag{10.6}$$

This definition of a continuous time diffusion can be found in several text books, such as Øksendal (1985).

The specification of the problem is completed by specifying the likelihood $p(y_k|x,t_k)$. In time and space-time, the problem is usually solved recursively by an updating formula. The updating formula can be cast in various forms. One form of the updating formula is (Miller *et al.* 1999)

$$p(x,t_k|y_k) = \frac{p(y_k|x,t_k)p(x,t_k|y_{k-1})}{p(y_k|y_{k-1})}. \tag{10.7}$$

Here, $p(x,t_k|y_{k-1})$ is the conditional density of the hidden process $X(t_k)$ given the observations up to the previous observation time t_{k-1} and $p(y_k|y_{k-1})$ is a constant.

In geophysical and environmental problems, the dimension of the state space may be very large, i.e. $m \approx 10^7$. Sequential estimation through the general formula (10.7) involves evaluation of complex integrals of similar or even higher dimensionality. Therefore, the general filtering techniques, for example through the use of Monte Carlo methods (Doucet *et al.* 2000) are computationally highly impractical. A further complication is due to the drift function $b(\cdot)$ being non-linear.

Gustafsson (2001) gives an overview of current use of statistical methods in weather forecasting, including data–assimilation and probabilistic forecasting techniques. The general approach to the filtering problem taken in weather forecasting seems to be a local-in-time linearization of $b(\cdot)$ and the assumption that all distributions involved are Gaussian. This allows for the use of extended Kalman filter techniques. Other approaches involve simulation from (10.4). Through the use of such *ensemble simulations*, one may estimate solutions to the Kolmogorov equation (10.5), see Evensen (1994) and Evensen (1997).

We may in some situations not be interested in an *a posteriori* estimate of the hidden process, but less ambitiously, a parameter of the process. In the case of European transboundary air pollution, our interest is in the national sulphur emissions, which enter linearly into a complex atmospheric transport-chemistry model. However, the physical model itself is in this case regarded as deterministic, since its code is not available to us.

10.3   Examples

Example 1: Conditional Analysis of European Air Pollution

Emitted sulphur and nitrogen compounds from different countries in Europe may be transported long distances and cause ecosystem damage when deposited. The main forms of impact are acidification and eutrophication. Tolerance limits for the ecosystems to these pollutants are called critical loads, and such critical loads have been calculated for most of Europe. This example concerns estimating geographical areas of critical load exceedance, while Example 2 will focus on estimating the emissions from each country in Europe.

By using reported national emissions inventories, the deposition of air pollutants is predicted on a 150 km × 150 km grid using an acid deposition model. An extensive monitoring network measures deposition at about 100 monitoring stations across Europe. Figure 10.1 shows the locations of stations monitoring wet deposition of nitrogen in 1995. We see that the density of monitoring stations is less dense in Southern and Eastern parts of Europe than in the Western parts.

Both emission inventories, atmospheric deposition modeling and monitoring activities are undertaken through the Co-operative Programme for Monitoring and Evaluation of the Long Range Transmission of Air Pollutants in Europe (the EMEP programme). The acid deposition model (AD–model) gives a smooth deposition field of expected or average deposition over 150 km × 150 km squares. In contrast, the measured depositions are specific to geographical point locations. The difference between these two sources of information may be due to local spatial variability, model errors or measurement errors. Figure 10.2 shows measured versus AD–model predicted deposition of nitrogen. We see that there is a tendency towards large errrors for large AD–model predictions. A log–transform was introduced in the modeling to ensure positive deposition estimates and stabilize error variance.

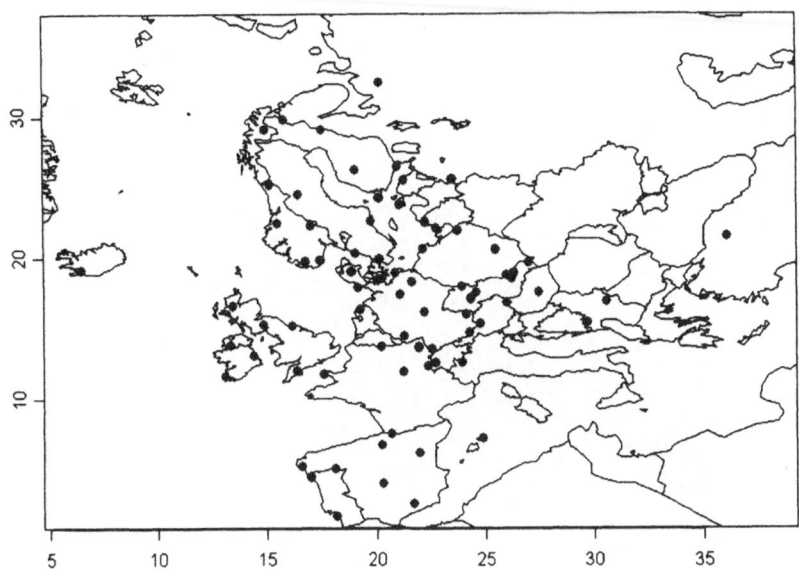

Fig. 10.1. Location of stations monitoring nitrogen deposition in 1995. Reprinted from *Atmospheric Environment*, volume 34, Hirst, D. *et al.*: Estimating the exceedance of critical loads in Europe by considering local variability in deposition, p 3790, Copyright (2000), with permission from Elsevier Science.

Of course, the extent of ecosystems is unrelated to the spatial resolution of the model grid. Furthermore, a smooth AD–model deposition field is inappropriate for assessing true exceedances of critical loads. Following Hirst *et al.* (2000), we suggest a statistical model for the true deposition field, combining the AD–model and the measurements. This is a model (on log-scale) of the type (10.4), where $Y(s,t)$ is the true deposition and x is the acid deposition model. The covariance is taken as

$$c(\|(s',t') - (s'',t'')\|; \theta) = c_s(\|s' - s''\|; \theta)\, \delta(t' - t''), \tag{10.8}$$

where time t is an index for year.

Denote by $l(s)$ the critical load at location s and by $I\{Y(s,t) > l(s)\}$ exceedance at s, where $I(\cdot)$ is the indicator function. The exceeded area for a region A at time t is

$$\alpha(A,t) = \int_A I\{Y(s,t) > l(s)\}\, ds. \tag{10.9}$$

Original

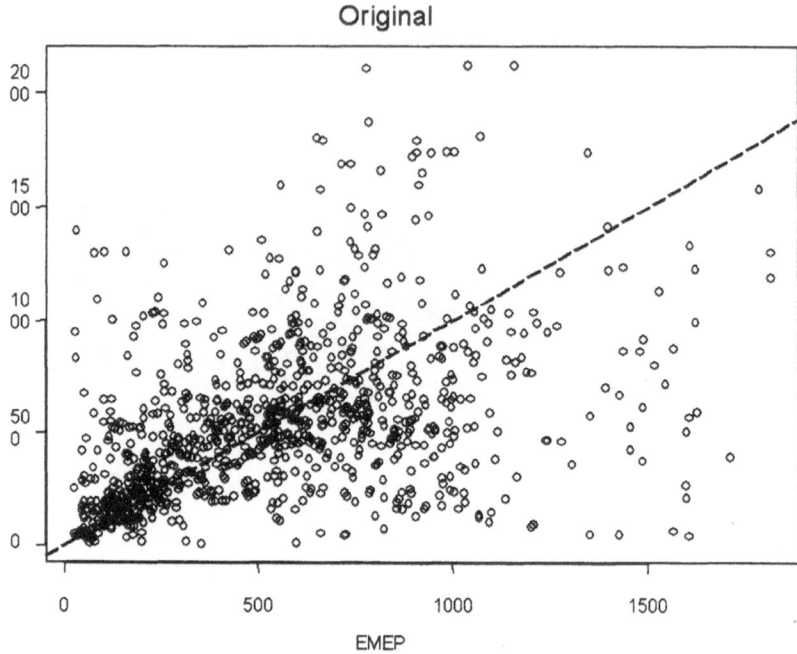

Fig. 10.2. Measured and AD–model predicted deposition of nitrogen during 1985–1995. Reprinted from *Atmospheric Environment*, volume 34, Hirst, D. *et al.*: Estimating the exceedance of critical loads in Europe by considering local variability in deposition, p 3792, Copyright (2000), with permission from Elsevier Science.

The probability distribution of the exceeded area $\alpha(A, t)$ is estimated by a plug-in method as follows. First, we estimate the spatial covariance parameters of the Y-field by weighted least squares. The fitted correlation function and empirical (bin-averaged) values for nitrogen are shown in Figure 10.3. Then we use conditional simulation (Ripley 1981) to generate realizations of the Y-field given the observations. From 1000 simulations of the Y-field we obtain 1000 realizations of $\alpha(A, t)$. A feature of the conditional simulation approach is that uncertainty due to the configuration of the monitoring network is accounted for, and this will be propagated into the exceeded area estimates.

Figure 10.4 shows time series of exceedance of critical loads for eutrophication during 1985-1995. The results show that the the estimated area of exceedance for Europe is larger for our method (full line) than for the method using only the AD-model predicted deposition (dotted line). The broken line shows the results using the conditional expectation (Kriging) of the deposition field. Such an approach would also fail to account for local spatial variability. The

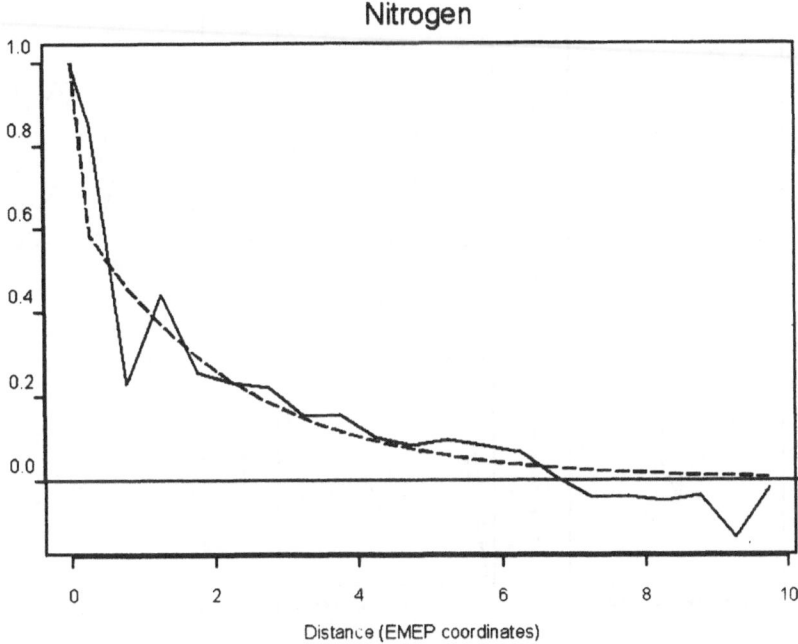

Fig. 10.3. Empirical correlation function (full line) and and fitted correlation function (broken line) for nitrogen. Each unit of distance corresponds to 150 km. Reprinted from *Atmospheric Environment*, volume 34, Hirst, D. *et al.*: Estimating the exceedance of critical loads in Europe by considering local variability in deposition, p 3793, Copyright (2000), with permission from Elsevier Science.

critical loads for acidification are for each location a bivariate function of both nitrogen and sulphur deposition. Extending the method to this situation requires joint modeling of nitrogen and sulphur space-time fields, as described by Hirst *et al.* (2000).

The AD–model is the key to describing the mean field of the deposition of air pollutants in Europe. However, it does not provide information on local spatial variability. Accounting for spatial variability is important for obtaining unbiased estimates of critical loads exceedance in Europe. Measured deposition in a monitoring network may provide this information. Thus, critical loads assessment is best done by combining these sources of information in a statistical model.

Our plug-in method does not account for uncertainty in the estimated parameters. This is best done in a Bayesian framework. Our next example illustrates Bayesian analysis of European air pollution, although in a slightly different context.

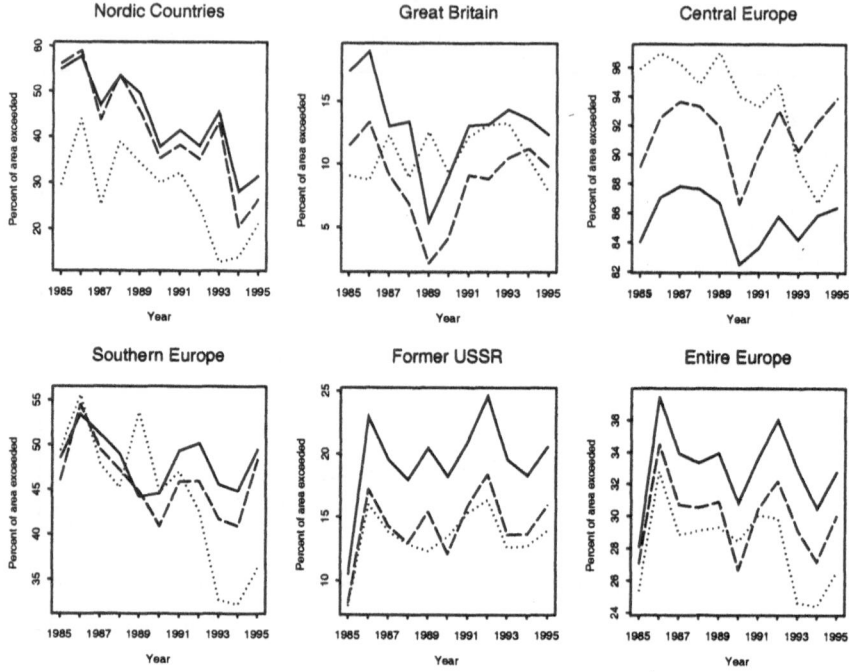

Fig. 10.4. Estimated critical load exceedance for nutrient nitrogen for various regions in Europe (full line). Also shown are estimates based on AD–model only (dotted line) and on Kriging (broken line). Reprinted from *Atmospheric Environment*, volume 34, Hirst, D. *et al.*: Estimating the exceedance of critical loads in Europe by considering local variability in deposition, p 3798, Copyright (2000), with permission from Elsevier Science.

Example 2: Bayesian Analysis of European Air Pollutant Emissions

The general set-up for this problem is the same as in Example 1, although the focus is different. Instead of estimating the effects, we turn our attention to the national emissions. The Geneva Convention of 1979 on Long-Range Transboundary Air Pollution and several follow-up protocols puts explicit obligations on the participating nations to reduce their emissions. By combining information from the previously described acid deposition model (AD–model) and monitored depositions, we may re-estimate the nationally submitted emissions inventories. In the future, one may wish to invoke such a tool for monitoring compliance with agreements on emissions reduction.

As described in Høst (1999a), we again use the statistical model (10.4), with minor modifications. First, the deposition predicted by the AD–model depends linearly on the emitted pollutant concentrations. We take the emissions as random, but the transport model is fixed. For a given year, we may omit the index for time and put

$$\widetilde{X}_h(s) = \widetilde{g}_h'(s)\,\beta, \qquad (10.10)$$

where $\widetilde{g}_h(s)$ denotes AD–model predicted deposition per unit emission and β is a vector of emissions from each country or region. We assume that the spatial covariance function may be parameterized by a variance σ^2 and a correlation range a parameter. As priors we use $p(\sigma^2) \propto \sigma^{-2}$ and $p(a) \propto (1 + a)^{-2}$, as suggested by Handcock & Wallis (1994). To complete the model, we need to specify a prior on β. Experts suggest taking the officially submitted values as prior mean with coefficient of variation 0.3. A priori, we take the national emissions to be independent normal, that is $\beta_j \sim N(\beta_{0j}, \gamma_0\beta_{0j})$. Now, the posterior is

$$p(\beta, \sigma^2, a|y) \propto p(\beta|\beta_0, \gamma_0)p(\sigma^2)p(a)L(\beta, \sigma^2, a|y) \qquad (10.11)$$

This posterior can be integrated by Markov Chain Monte Carlo (MCMC) methods to give the posterior mean. A computationally much cheaper alternative, which is followed here, is to maximize (10.11) to give the posterior mode. An estimate of the covariance matrix of the posterior mode is provided by the inverse Hessian of the log posterior.

The results for sulphur emissions in 1990 is shown in Table 10.1 for some of the large emittors. We see that the the posterior estimates differ from the prior emissions, in particular for the United Kingdom, Italy and the former German Democratic Republic. For the German Democratic Republic, the posterior estimated emission is lower than what was officially reported. This seems at first counterintuitive. However, considering the political situation around 1990, it is likely that industrial activity was quite low during this year, a fact that may not have been picked up in the official emission reports. Our method also give uncertainty estimates in terms of posterior CV. The CV goes down somewhat, in particular for the United Kingdom and Italy, but less so for Ukraine. This is an effect of the configuration of the monitoring network, since there are very few stations that are able to capture deposition transported from Ukraine. The statistical model provides a framework for assessing revision of the monitoring network, optimized towards monitoring national compliance with obligations to reduce emissions. The optimal configuration of a compliance monitoring network is difficult to assess intuitively and should be solved by numerical optimization techniques.

Both the issues of emissions estimation and monitoring network design are problems that cannot be solved satisfactory without combining a physically-based model with a statistical modeling approach. The AD–model provides the explanatory tool for relating emissions to depositions, although on a coarse level. The statistical model explains the deviations between observed deposition and AD–model predictions, thereby quantifying uncertainty. By

Table 10.1. Results of emissions estimation. Prior Emission and Posterior Estimated Emission are in units of 1000 tons of sulphur dioxide

Country	Reported	Estimated	Rel. Change	CV
German Democratic Republic	4755	4545	0.96	0.28
Italy	2251	2715	1.21	0.24
Poland	3210	3111	0.97	0.21
United Kingdom	3760	4206	1.12	0.21
Ukraine	3850	3904	1.01	0.30

allowing for the attribution of some of the observed uncertainty to the emissions, the statistical model also allows for re-estimation and checking of reported national emissions.

Example 3: Conditional Analysis of Norwegian River Runoff

This example concerns a flood forecasting problem where we have available information from two physical models, as well as measurements. The purpose of the statistical analysis is to provide error estimates for the flood forecasts. The forecasting system is run routinely by the Norwegian Water Resources and Energy Administration. Numerical forecasts of temperature and precipitation up to six days ahead are produced by a numerical weather forecasting model (WF–model) by the Norwegian Institute of Meteorology. The WF–model is based on combining information from a global weather forecasting model and a high resolution limited area forecasting model. Forecasted values of precipitation and temperature are input to the hydrological runoff model (R–model). The R–model is an implementation of the Swedish HBV–model (Bergström & Forsman 1973), which is widely used throughout Scandinavia. It contains a quantitative description of important physical processes such as snow accumulation, melting, soil moisture, ground water dynamics and routing of the runoff through the catchment. The combination of the WF–model and the R–model produces river runoff forecasts up to six days ahead for selected catchments. There are measurements available on both meteorological and hydrological variables. The goal is to provide uncertainty estimates, not to improve the models. Therefore, the problem is one of conditional analysis. Denote by x_1 the weather forecast and by y_1 the weather observations. Similarly, denote by x_2 the R–model predicted runoff and by y_2 the runoff measurements. We are interested in the probability density of y_2 given both the WF– and H– model forecasts x_1, x_2 and the past history. This density may be written

$$p(y_2|x_1,x_2,past) = \int p(y_2|y_1,x_2,past)\, p(y_1|x_1,past)\, dy_1. \quad (10.12)$$

Here, the joint density $p(y_1,y_2|x_1,x_2,past)$ has been decomposed into densities $p(y_2|y_1,x_2,past)$ and $p(y_1|x_1,past)$ on the right hand side of (10.12).

The density $p(y_2|y_1, x_2, past)$ describes the uncertainty of the runoff predicted by the R–model for various weather regimes, while $p(y_1|x_1, past)$ describes the uncertainty of the weather variables forecasted by the WF–model. Although $p(y_2|x_1, x_2, past)$ could be modelled directly without using the meteorological data, an advantage of our decomposition approach is that we may quantify the contribution from each sub–model to the total flood forecast error.

The specific modeling now involves quantification of weather forecast errors and flood prediction errors. An overview is presented here; for details the interested reader is referred to Langsrud *et al.* (1998a), Follestad & Høst (1998) and Langsrud *et al.* (1998b). The weather forecast error is described by $p(y_1|x_1, past)$, where x_1 and y_1 are bivariate time series of precipitation and temperature. A detailed data analysis suggests that the observed temperature may be fitted to a Gaussian distribution through regression on the forecast and past values. Similarly, the observed precipitation occurrence was fitted to a logit model and conditionally positive precipitation was fitted to a gamma distribution. The logit–gamma statistical model for precipitation was originally suggested by Stern & Coe (1984). An example of the fit to a conditional gamma distribution for the Knappom catchment in Eastern Norway is shown in Figure 10.5. We see that the fit is very good. By simulating from the fitted statistical model, we may calculate confidence intervals for the WF–model. An example resulting from 1000 Monte Carlo simulations is shown in Figure 10.6. We see that the correspondence between observed and forecast precipitation is not very good; in fact the weather forecasting model tends to overestimate the probability of precipitation occurrence and has problems predicting the correct amounts. This is due to the inherent difficulties in providing deterministic forecasts of local precipitation amounts. The estimated upper confidence limit for precipitation shows less detail for forecasts with longer lead times, and suggests that for 6 days ahead forecasts the upper confidence limit is 10-20 mm irrespective of the forecast.

The analysis of runoff from the R–model and observations leads us to work with observed–predicted runoff residuals on the log–scale. These residuals were fitted to an AR(1) model with coefficients and noise variance depending on R–model predictions, past values and regimes of snow cover, precipitation and temperature. An example of the R–model predictions and fitted R–model error for the Knappom catchment is shown in Figure 10.7. We see that the fitted R–model error median (0.5–percentile) follows the observed runoff closer than the R–model predictions. Thus, we may improve on the R–model by using the fitted statistical model for calibration. Although such a calibration may give more predictive power, it may reduce interpretability of the forecasts. However, statistics in this case proves to be an important diagnostic tool for the runoff model.

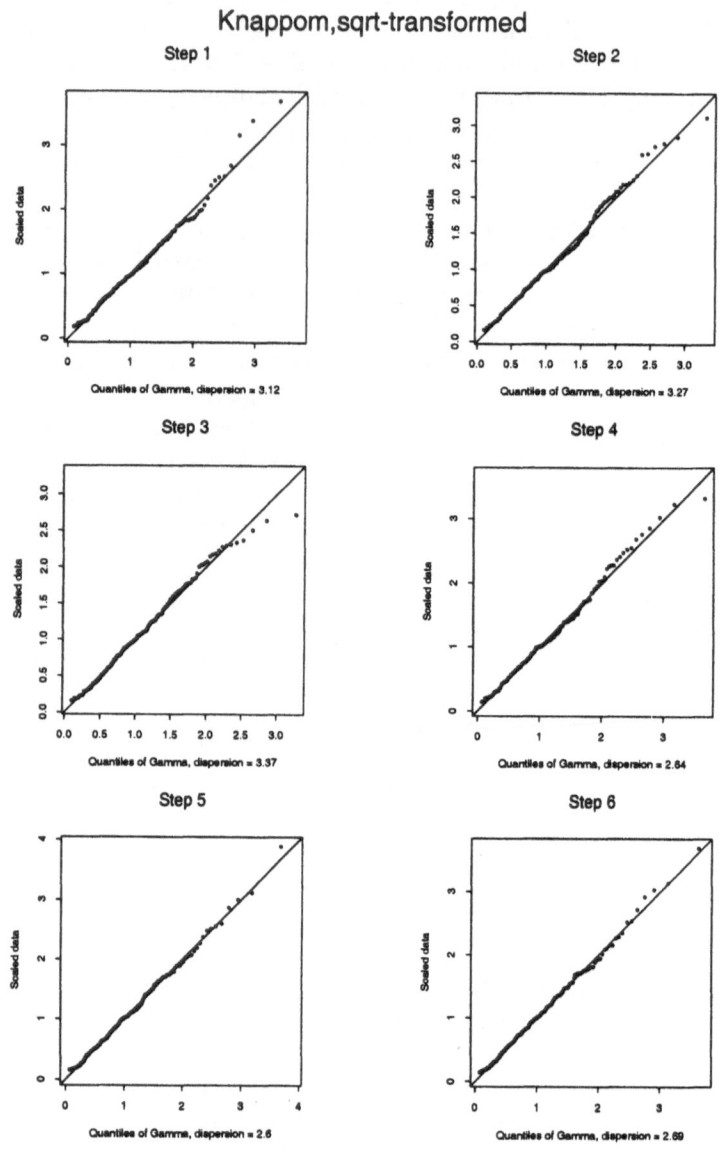

Fig. 10.5. A fit of precipitation amounts to the gamma distribution.

Figure 10.8 shows the forecast uncertainty of the combined meteorological-hydrological runoff forecasts. We see that the forecast uncertainty increases substantially with forecast lead time.

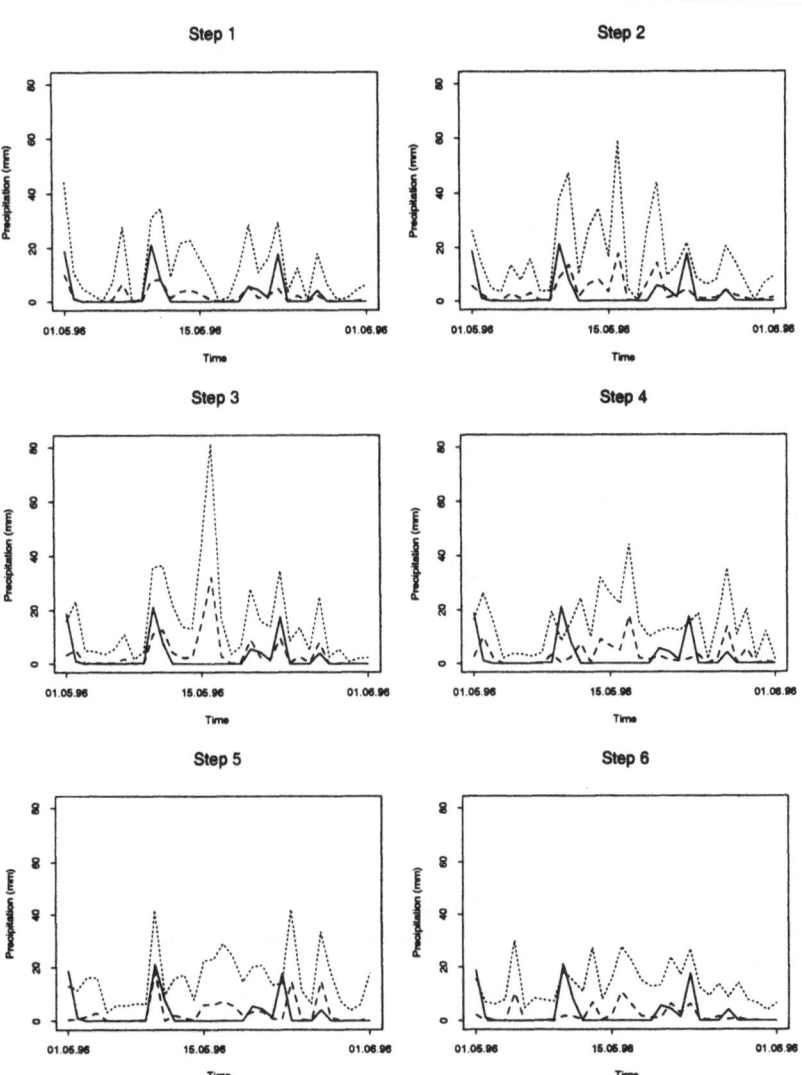

Fig. 10.6. Precipitation model forecasts (dashed lines), observations (full lines) and estimated upper 95% confidence bounds (dotted lines).

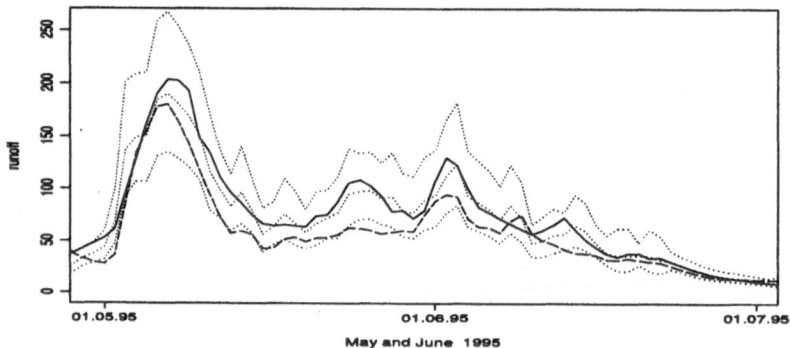

Fig. 10.7. Runoff model prediction (broken line), observed runoff (full line) and 0.025–, 0.5– and 0.975–percentiles of the fitted model error distribution.

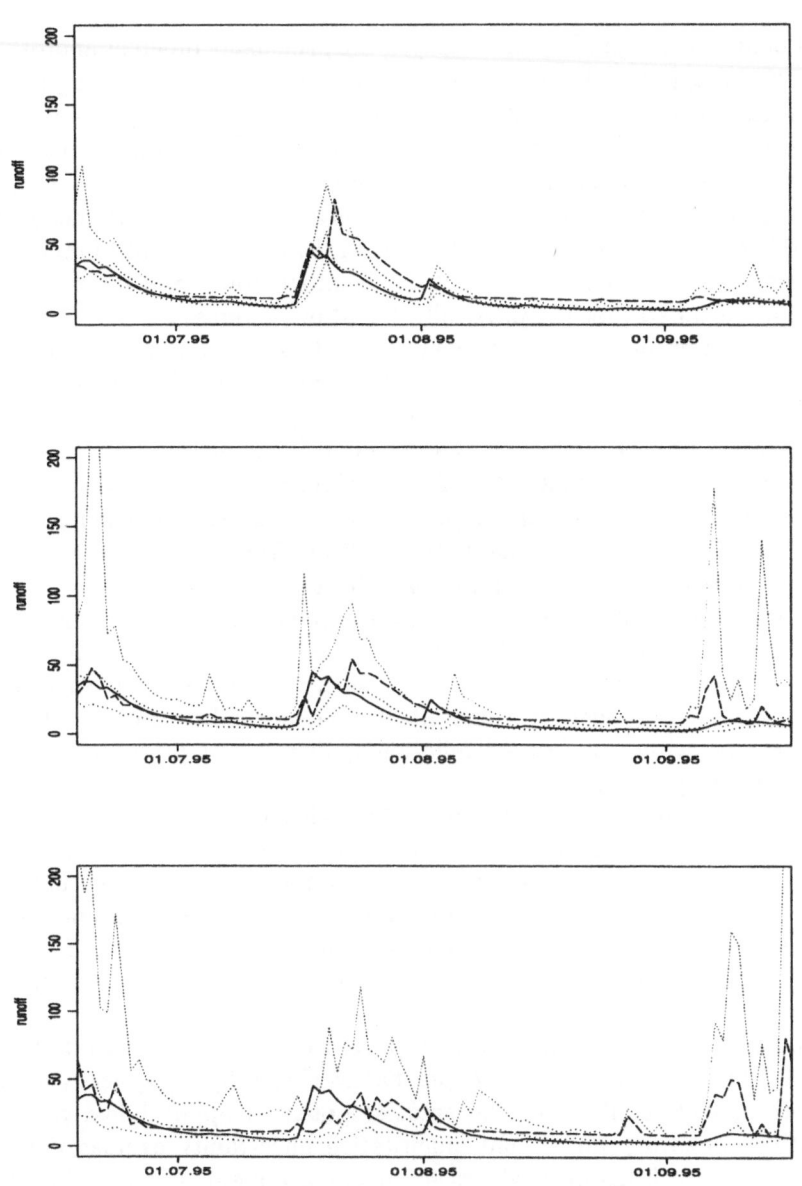

Fig. 10.8. Runoff forecasts based on meteo–hydrological models for 1– (upper panel), 3– (middle panel) and 6– (lower panel) day ahead forecasts. Forecast (broken line), observed runoff (full line) and 0.05–, 0.5– and 0.95– percentiles of the fitted forecast error distribution.

10.4   Discussion

Conditional analysis and Bayesian analysis are principally different approaches to combining physical and statistical modeling. Bayesian modeling is the natural framework for doing statistical analysis within the physical model, i.e. internal analysis. In contrast, in a conditional analysis a statistical model is built external to the physical model. Historically, the impact of physically–based explanatory tools in science and applications has been tremendous. By working inside the physical model, the statistician may make important contributions to optimize the model's explanatory power and by providing uncertainty estimates for the physically–based predictions.

In contrast, conditional analysis is the work mode of choice when the physical model is inaccessible for the statistician. This line of work also has some distinct advantages. Conditional analysis involves very few assumptions about the physical model. Therefore, it is well-suited for model criticism and for diagnostic purposes. In contrast, a Bayesian approach involves the specification of prior distributions and hyperparametrers, possibly through several hierarchical levels. Distinguishing between effects from a complex nonlinear physical model and prior assumptions may be a cumbersome task.

Deviations between observations and physical model predictions may result from measurement error, model error and small–scale variability. Measurement error is often quite easy to describe, since it tends to be uncorrelated in space and time. Model error and small–scale variability is often related, due to non-linearities. In data assimilation applications, a prior model error needs to be prespecified.

For many practical applications, a detailed description of small-scale variability is important. Examples include assessment of critical loads exceedance and local climate change. Modeling small-scale variability directly from physics is extremely complicated due to the closure problem for the non-linear differential equations involved. To this end, an empirical approach, or the semi-empirical approach described in this article, may be the most practical approach.

An important question, related to the topic of this paper is whether a physical model should be deterministic or stochastic. Generally, a physically based stochastic model should have the greatest potential applicability, providing both explanatory power and quantitative uncertainty estimates. Furthermore, such a framework allows for drawing on well–known principles of likelihood–based inference. Within the geophysical sciences the work done within data assimilation is a step in this direction. Data assimilation may be regarded as an application of Bayesian non–linear filtering. The approach taken within this community seems to be to retain as much detail as possible in the physical model, by allocating computer resources mainly to solving the dynamics on the finest possible grid. This results in rather coarse error estimates. In contrast, by using less physical details, one may afford to estimate full posterior distributions. This could be very useful for assessment of extreme events and exceedance of critical levels.

I have briefly described two principally different classes of approaches to combining physical and statistical modeling. The real challenge is maybe better described as combining statistical and physical thinking into the model building process. To build a model that can balance quantitatively physical knowledge with available data for a complex phenomenon is an extremely challenging research task.

References

Bergström, S. & Forsman, A. (1973). Development of a conceptual deterministic rainfall-runoff model, *Nordic Hydrology*.

Cox, L. & Piegorsch, W. (1996). Combining environmental information. 1. Environmental monitoring, measurement and assessment. *Environmetrics* **7**(3), 299–308.

Doucet, A., Godsill, S. & Andrieu, C. (2000). On sequential Monte Carlo sampling methods for Bayesian filtering. *Statistics and Computing* **10**(3), 197–208.

Evensen, G. (1994). Sequential data assimilation with a nonlinear quasigeostrophic model using Monte-Carlo methods to forecast error statistics. *Journal of Geophysical Research – Oceans* **99**(C5), 10143–10162.

Evensen, G. (1997). Advanced data assimilation for strongly nonlinear dynamics. *Monthly Weather Review* **125**(6), 1342–1354.

Follestad, T. & Høst, G. (1998). A statistical model for the uncertainty in meteorological foreecasts, with applications to the Knappom and Røykenes catchments. Techical Report No. SAMBA/16/98, Norwegian Computing Center, Oslo, Norway.

Gustafsson, N. (2001). Statistical issues in weather forecasting. *Scandinavian Journal of Statistics*, to appear.

Hirst, D., Kåresen, K., Høst, G. & Posch, M. (2000). Estimating the exceedance of critical loads in Europe by considering local variability in deposition. *Atmospheric Environment* **34**(22), 3789–3800.

Høst, G. (1999). Bayesian estimation of European sulphur emissions using monitoring data and an acid deposition model. *Environmental and Ecological Statistics* **6**(4), 381–399.

Langsrud, Ø., Frigessi, A. & Høst, G. (1998). Pure model error for the HBV-model. Techical Report No. SAMBA/05/98, Norwegian Computing Center, Oslo, Norway.

Langsrud, Ø., Høst, G., Follestad, T., Frigessi, A. & Hirst, D. (1998). Quantifying uncertainty in HBV runoff forecasts by stochastic simulations. Techical Report No. SAMBA/20/98, Norwegian Computing Center, Oslo, Norway.

Miller, R., Carter, E. & Blue, S. (1999). Data assimilation into nonlinear stochastic models. *Tellus Series A - Dynamic Meteorology and Oceanography* **51**(2), 167–194.

Murphy, A. (1995). A coherent method of stratification within a general framework for forecast verification. *Monthly Weather Review* **123**(5), 1582–1588.

Øksendal, B. (1985). *Stochastic Differential Equations. An Introduction with Applications. First Edition*, Springer-Verlag.

Olkin, I. (1996). Meta-analysis: Current issues in research synthesis. *Statistics in Medicine* **15**(12), 1253–1257.

Piegorsch, W. & Cox, L. (1996). Combining environmental information. 2. Environmental epidemiology and toxicology. *Environmetrics* **7**(3), 309–324.

Ripley, B. (1981). *Spatial Statistics*. Wiley, New York.

Stern, R. & Coe, R. (1984). A model fitting analysis for daily rainfall data. *Journal of the Royal Statistical Society, Series A* **147**(1), 1–34.

Tritton, D. (1977). *Physical Fluid Dynamics*. Van Nostrand Reihold, Molly Millars Lane, Wokingham, Berkshire, UK.

11   Bayesian Analysis of Computer Code Outputs

Marc C. Kennedy[1], Anthony O'Hagan[2], and and Neil Higgins[3]

[1] National Institute of Statistical Sciences, PO Box 14006, Research Triangle Park, North Carolina, NC 27709-4006, USA
[2] Department of Probability & Statistics, University of Sheffield, Sheffield S3 7RH, UK
[3] National Radiological Protection Board, Chilton, Didcott, Oxfordshire OX11 0RQ, UK

Abstract. Complex computer models are widely used to describe and predict environmental phenomena. Although such models are generally deterministic, there is a flourishing area of research that treats their outputs as random quantities, in order to provide powerful solutions to some important problems facing the users of such models. Such problems include interpolation/emulation of the computer code itself, sensitivity and uncertainty analysis, and calibration.

This article reviews this field, with particular reference to a Bayesian methodology that offers a unified framework for addressing all such problems. A substantial practical example is then presented, involving calibration of a model to describe the radioactive deposition following the Windscale nuclear accident in the UK in 1957. The example illustrates some important features of the approach when the computer model is unable to represent the real phenomenon accurately, so that the Bayesian method then seeks to correct the computer model.

11.1   Analysis of Computer Code Outputs

11.1.1   Deterministic Computer Models

Computer models are used throughout science and technology to describe complex real-world phenomena. They are typically used to predict the corresponding real-world phenomenon, as in the following examples.

- Modern weather forecasting is done using enormously complex models of the atmosphere (and its interactions with land and sea). The primary intention is to predict future weather, given information about current conditions.
- Manufacturers of motor car engines build models to predict their behaviour. They are used to explore possible variations in engine design, and thereby to avoid the time and expense of actually building many unsuccessful variants in the search for an improved design.
- Water engineers build network flow models of sewer systems, in order to predict where problems of surcharging and flooding will arise under rainstorm conditions. They are then used to explore changes to the network to solve those problems.

- Models of atmospheric dispersion are used to predict the spread and deposition of chemicals or radionuclides released from industrial or nuclear installations. They are used in assessing the risks of such installations, during normal operation, in the event of hypothetical accidents, and in supporting the emergency response to actual incidents.

All of these models, and many more, have implications for the environment, whether in terms of determining when it is safe to apply pesticides to a crop, reducing motor car emissions, avoiding sewage discharges into rivers, or understanding the extent of pollution following accidents. The methodology presented here is applicable to such models generally. It is illustrated through an example of the last kind of model, an atmospheric dispersion model applied to the Windscale nuclear accident in the North-West of England in 1957.

Such models are usually deterministic, in the sense that if the computer code is run a second time with the same inputs then it will produce exactly the same outputs. Nevertheless, there is a flourishing area of study that treats these outputs as random quantities, in order to provide powerful solutions to some important problems facing the users of such models. This field is known by the acronym SACCO, the Statistical Analysis of Computer Code Outputs.

11.1.2   DACE

In order to understand the paradox of statistical analysis of deterministic computer code outputs, it is helpful to begin with the precursor of SACCO, in the area known as DACE, the Design and Analysis of Computer Experiments. Consider the engine simulator described above. Such computer codes can take hours to run for a single configuration of inputs that describe a particular engine design. The designers wish to vary these inputs so as to determine a design to optimise some criterion. We can think of the value of this criterion for a given vector x of inputs as a function $\eta(x)$, or more generally we can think of $\eta(x)$ as any single output or function of outputs of interest. The task of optimisation is to find x to maximise or minimise $\eta(x)$.

Formulated in this way, the problem is simply the familiar task in numerical analysis of optimising a function of several variables. Various algorithms exist to search for the solution. These methods involve repeatedly evaluating $\eta(x)$ at a series of input configurations that converges eventually to the solution. However, the long run times and the large number of inputs involved mean that this search is either an extremely time-consuming task or even simply impractical.

The idea of using statistical methods for this problem begins with thinking about applying regression methods to fit a response surface that approximates the function $\eta(.)$. That is, we have a set of evaluations $y_i = \eta(x_i)$, $i = 1, 2, \ldots, n$, which we treat as data (y_i, x_i) for a conventional regression analysis. The y_is are not really random variables, and it is clear that replication is not possible since running the code again with the same inputs produces the same output. There is no observation error, yet these methods will clearly be useful in guiding the search for an optimal engine design. We can think of the set of computer runs as a 'computer experiment' and their outputs as the data from such an experiment.

A key step in this development was to move to a nonparametric framework, which allowed fitting of an arbitrary function $\eta(\cdot)$ to the data, and simultaneously allowed the recognition of the fact that there is no observation error. In this approach, $\eta(\cdot)$ is modelled as a random function that is a realisation from an appropriate stochastic process over the x space. Suppose that we model $\eta(\cdot)$ via

$$E(\eta(\boldsymbol{x})) = \boldsymbol{h}(\boldsymbol{x})^T \boldsymbol{\beta}, \qquad \mathrm{Cov}(\eta(\boldsymbol{x}), \eta(\boldsymbol{x}')) = \sigma^2 r(|\boldsymbol{x} - \boldsymbol{x}'|),$$

where $\boldsymbol{h}(\cdot)$ is a vector of known regression functions, $\boldsymbol{\beta}$ and σ^2 are unknown parameters and $r(\cdot)$ is a known correlation function. The latter has the property that $r(0) = 1$ and $r(d)$ decreases as d increases. This expresses the fact that if two input configurations $\boldsymbol{x}$ and $\boldsymbol{x}'$ are close together then the outputs $\eta(\boldsymbol{x})$ and $\eta(\boldsymbol{x}')$ will be highly correlated, but that the correlation decreases as the input configurations move further apart.

It is then possible to construct a best linear unbiased estimator of the output $\eta(\boldsymbol{x})$ at any untried input configuration $\boldsymbol{x}$ using the data $y_1, y_2, \ldots, y_n$. The primary practical issues arising in this theory are the choice (and estimation from the data) of the correlation function $r(\cdot)$ and the choice of input configurations $\boldsymbol{x}_1, \boldsymbol{x}_2, \ldots \boldsymbol{x}_n$ at which to run the code to provide optimal information about $\eta(\cdot)$. The former problem is very closely related to the choice of a spatial correlation function in the field of geostatistics (also known as 'kriging'), although the latter is concerned primarily with functions of two or three variables while DACE will often demand much higher dimensional input spaces. The problem of the choice of $\boldsymbol{x}_i$s is one of experimental design. An important review of DACE work is Sacks et al. (1989), and for more recent work on computer experiment design see Bates et al.(1996). A good reference for geostatistics is Stein (1999).

11.1.3   SACCO

The main focus in DACE was to predict the output $\eta(\boldsymbol{x})$ of the computer code at some untried input configuration. The result is a statistical approximation to the code, that agrees with the observed outputs at the design inputs but interpolates smoothly between them to allow prediction at other inputs. The approximation is called an *emulator*. Its value arises from the fact that it is much faster than the original simulator code, since its predictions are almost instantaneous, while a good design may allow it to be a very good approximation to the original code. Prediction/interpolation is just one of a number of problems that concern users of computer models.

SACCO can be seen as the use of statistical methods more generally to address the whole range of problems concerning computer codes. A SACCO workshop was held in Wales with support from the EPSRC in 2000, and general information may be obtained from the SACCO website [`www.maths.dur.ac.uk/SACCO`]. The principal questions facing computer code users are as follows.

1. *Interpolation* — inference about $\eta(x)$ for an untried input x. As discussed, this problem and extensions such as optimising $\eta(x)$ with respect to x have been extensively covered in the DACE literature.

2. *Sensitivity analysis* — study of how individual components of x influence $\eta(x)$. This can be done in the local sense of studying derivatives of $\eta(\cdot)$ or in the more global sense of considering varying inputs over some range. If distributions are assigned to the components of x, we can average with respect to those distributions in order to obtain sensitivity measures. Yet another acronym is used for this sub-area, SAMO, the Sensitivity Analysis of Model Outputs. Some international conferences have been run under that title in recent years; a key reference is Saltelli *et al* (2000).

3. *Uncertainty analysis* — inference about the uncertainty in $\eta(x)$ that is induced by uncertainty about x. It is common for computer codes to require as inputs physical parameters and other factors that are not known. In order to use the code for prediction, it is necessary to estimate these inputs, but simply to use the resulting output as a prediction ignores the uncertainty induced by not knowing the estimated inputs. To the extent that uncertainty analysis involves placing a distribution over the uncertain inputs, there are clear connections with sensitivity analysis, but the objectives of the analysis are somewhat different.

4. *Calibration* — using observations of the real-world phenomenon to make inference about uncertain input parameters. Where it is necessary to provide estimates of unknown input parameters, this is often done using expert judgement, but where there are observations available of the real-world phenomenon these can be used to learn about those parameters. Calibration is traditionally seen as a process of 'fitting' the computer model to the observational data, by adjusting the uncertain inputs until the model predicts those data as closely as possible.

5. *Model uncertainty* — quantifying the magnitude of possible discrepancies between the computer model and reality. Any model is necessarily an imperfect approximation to the reality that it represents. Errors in the structure of the model itself, due for instance to imperfect understanding of the underlying science, contribute to these imperfections (as do bugs in the computer code). There is considerable interest among users of computer codes in their accuracy and validity. Where observational data are available, these can be used to estimate model inadequacies and to quantify the model uncertainty through, for instance, an error variance.

Another approach is to analyse the model itself and to consider sensitivity or uncertainty analysis of variations in the model structure.

11.2   Bayesian Methods

11.2.1   Overview

Although SACCO encompasses other approaches, its most powerful techniques are based on extensions of the DACE methodology. In particular, we consider here a Bayesian approach that allows all the problems described in Section 11.1.3 to be addressed within a unified framework. The key step is already implicit in the DACE work; we represent $\eta(\cdot)$ as a random function. Within the Bayesian approach, this formally represents a prior distribution for $\eta(\cdot)$. Combining this with the data from runs of the code yields a posterior distribution.

Despite the simplicity of this conception, its implementation introduces a number of important technical issues. The most detailed description of these is in Kennedy and O'Hagan (2001).

The earliest references to Bayesian analysis of computer code outputs are within the DACE literature, where the Bayesian approach is exploited in the context of interpolation. The Bayesian formulation is mentioned in Sacks *et al* (1989), and more recent references are Currin *et al* (1991) and Morris *et al* (1993).

Problems of sensitivity and uncertainty analysis, where a distribution is placed on x, have typically been addressed in the past by Monte Carlo methods. That is, the x_is are random draws from that distribution, so that the y_is are then random draws from the uncertainty distribution of the output. Monte Carlo methods are appropriate when the computer code is not too complex, so that thousands of runs can be made in a reasonable time. However, by modelling $\eta(\cdot)$ as a random function, we can take advantage of the information that the output at two points with similar inputs should also be similar. This property that the function $\eta(\cdot)$ is smooth in some sense is not used by the Monte Carlo method. Accordingly, a Bayesian analysis can produce useful inferences from far fewer code runs. This is demonstrated for the problem of uncertainty analysis by Haylock and O'Hagan (1996), Oakley and O'Hagan (2000). A Bayesian approach to sensitivity analysis is briefly considered in O'Hagan *et al* (1999), and a more detailed paper is in preparation.

The first Bayesian approach to calibration is given by Craig *et al* (1996), employing Bayes linear methods, and this is extended to a fuller treatment of calibration and model uncertainty in Craig *et al* (2001). A fully Bayesian treatment is given in Kennedy and O'Hagan (2001). We will now present an outline of that analysis and a new application that illustrates some important issues.

11.2.2   Bayesian Calibration

Our model comprises the following components. We have two types of data, the first of which is the outputs from running the computer code N times,

$$y_j = \eta(x_j^*, t_j), \quad j = 1, 2, \ldots, N.$$

Notice that we now divide the inputs for the computer model into two parts. The vector x_j^* contains the values of the variable inputs for the j-th code run, while t_j contains the values of the calibration inputs. Remember that for calibration we wish to learn about the values of some unknown input parameters, which we call the calibration parameters. These have some true but unknown values θ. When we run the computer code, we have to specify values for these inputs in order to run the model and obtain outputs, so the t_j represents the input values specified for those calibration parameters on the j-th code run.

The variable inputs are other inputs, whose meaning is perhaps best seen by considering the second type of data. We have n observations of the real-world phenomenon. The variable inputs identify the context of each of these observations, so that the i-th observation is characterised by having variable inputs x_i. In the example to be presented in Section 11.3, the model describes deposition of radionuclides after a release of radioactive material into the atmosphere. The observations are measurements of deposition at various geographical locations, so the variable inputs serve to specify the location of each observation. We can consider them as latitude and longitude, for instance. If the model also had a time element, then it would be important to specify the time at which each measurement was made, and this would become a third variable input.

The observations are written

$$z_i = \zeta(x_i) + \varepsilon_i, \quad i = 1, 2, \ldots, n.$$

The terms ε_i denote observation errors, assumed to be iid $N(0, \lambda)$. The function $\zeta(\cdot)$ denotes the real-world process that the computer code $\eta(\cdot, \cdot)$ is modelling. Notice that in order to identify a real-world process value it is only necessary to specify the variable input values, since the unknown calibration parameters θ are fixed throughout.

We now need to relate the real process $\zeta(\cdot)$ to the model $\eta(\cdot, \cdot)$ which we do via the equation

$$\zeta(x) = \rho\,\eta(x, \theta) + \delta(x).$$

This shows the role of the true calibration parameters in linking the two functions. There are two more elements of this equation which represent ways in which the computer model fails to represent reality correctly. First, the scaling parameter ρ is used to allow for a systematic discrepancy between the code and reality. It may be that $\eta(\cdot, \theta)$ is broadly of the same shape as

$\zeta(.)$, but varies either too much or too little, and this can be accounted for by a value of ρ below or above 1.

The function $\delta(\cdot)$ is the model inadequacy function, which explains the remaining discrepancies between $\zeta(\cdot)$ and $\rho\eta(\cdot,\boldsymbol{\theta})$. Modelling the relationship between the computer code and reality through both ρ and $\delta(\cdot)$ in this way has been found to be useful in applications. Explicitly incorporating such a relationship is essential to a proper representation of the calibration problem, but it can have some unexpected consequences, as the example in Section 11.3 demonstrates.

Both $\eta(\cdot,\cdot)$ and $\delta(\cdot)$ are treated as unknown functions, for which we therefore need to specify prior information. We assume independent Gaussian process prior distributions for these functions. The use of Gaussian processes in this way is explained fully in Kennedy and O'Hagan (2001). Prior distributions are also required for $\boldsymbol{\theta}$, ρ, λ and various hyperparameters introduced in the Gaussian process modelling.

Although the obvious subject for inference in a calibration problem is the calibration parameter vector $\boldsymbol{\theta}$, the ultimate objective is usually to predict the real-world process $\zeta(\boldsymbol{x})$ at some unobserved variable input configuration $\boldsymbol{x}$.

The traditional approach to calibration consists of adjusting the calibration inputs in an *ad hoc* or systematic way to minimise a measure such as the sum of squares of the residuals $z_i - \eta(\boldsymbol{x}_i, \hat{\boldsymbol{\theta}})$, and then predicting $\zeta(\boldsymbol{x})$ by $\eta(\boldsymbol{x}, \hat{\boldsymbol{\theta}})$. There are two key features of the Bayesian analysis described in Kennedy and O'Hagan (2001) which improve on this traditional approach.

- The Bayesian prediction can take full account of all the relevant uncertainties. This is achieved by integrating the posterior distribution with respect to all the parameters, and making inference from the marginal posterior distribution of $\zeta(\boldsymbol{x})$ given just the data $y_1, y_2, \ldots, y_N, z_1, z_2, \ldots, z_n$. In particular, this analysis accounts for the remaining uncertainty in the calibration parameters $\boldsymbol{\theta}$ after calibration, whereas the traditional method simply uses a 'plug-in estimation' approach.
- The Bayesian prediction also estimates the scaling parameter ρ and the model inadequacy $\delta(\boldsymbol{x})$. In effect, it tries to *correct* the discrepancy between the code and reality.

In practice, although it is technically possible to integrate out all the parameters, Kennedy and O'Hagan (2001) advocate not integrating over some of the hyperparameters (adopting instead a 'plug-in' approach with regard to these), on the grounds that the uncertainty in these is sufficiently small to be ignored. The major uncertainties mentioned above, with respect to $\boldsymbol{\theta}$, $\eta(\cdot,\cdot)$ and $\delta(\cdot)$, are all fully accounted for in the analysis.

11.3   Windscale Nuclear Accident

11.3.1   Background

The National Radiological Protection Board (NRPB) develops advice and guidance on the protection of the public in the event of a nuclear accident in the UK (NRPB 1990a, b). This guidance is applied by organisations and agencies in the development of emergency plans, and would be used to help decision makers directly in the event of an accident. An important aspect of the guidance and legislative framework on accident management stems from the need to act quickly to minimise any exposure of the public to radiation. The guidance therefore deals with decisions concerning sheltering and evacuation, and EU legislation deals with restrictions on the sale of food. To enable a decision maker to deal more effectively in the future with decisions that must be taken quickly NRPB recognises the importance of being able to assess the situation at a very early stage. Various types of data will normally be available to help in this task, but in this paper we restrict our attention to new approaches to the assessment of radioactive deposition. We should emphasise that the work presented here is not intended as in any sense a definitive analysis of the Windscale accident, but is indicative of what can be achieved by the Bayesian approach.

Physical measurements of ground deposition may require laboratory analysis and are typically very limited in the early stages of an accident. An additional problem is that predictions are required over a wide area, but data might only be available in regions which are accessible by road, or close to the release source at fixed monitoring sites. We must also take into account the error associated with the measurement process.

Our data are from a reactor fire which started on the 10th October 1957 at Windscale in Northwest England. Detailed descriptions of the accident can be found in Arnold (1992). The accident is the most serious of its kind to have occurred in the UK. The NRPB has recently created a computerised database from the original data, part of which is used here.

Weather conditions during the release were very variable, due to the passage of a cold front over the region. The wind direction changed many times, resulting in a very non-gaussian pattern of deposition. Figure 1 shows the locations of the data points, which are measured depositions of iodine-131 (^{131}I). The grey-scale represents an interpolation of log-deposition at these points using S-Plus. The source location is marked as 'S'.

Notice that the data points are poorly distributed within the prediction region of interest, with many of the measurement points clustered close to the accident site.

11.3.2   Gaussian Plume Model

A simple computer code $\eta(x, t)$ is used to predict deposition, based on a Gaussian plume model (Clarke 1979) with plume depletion. The variable

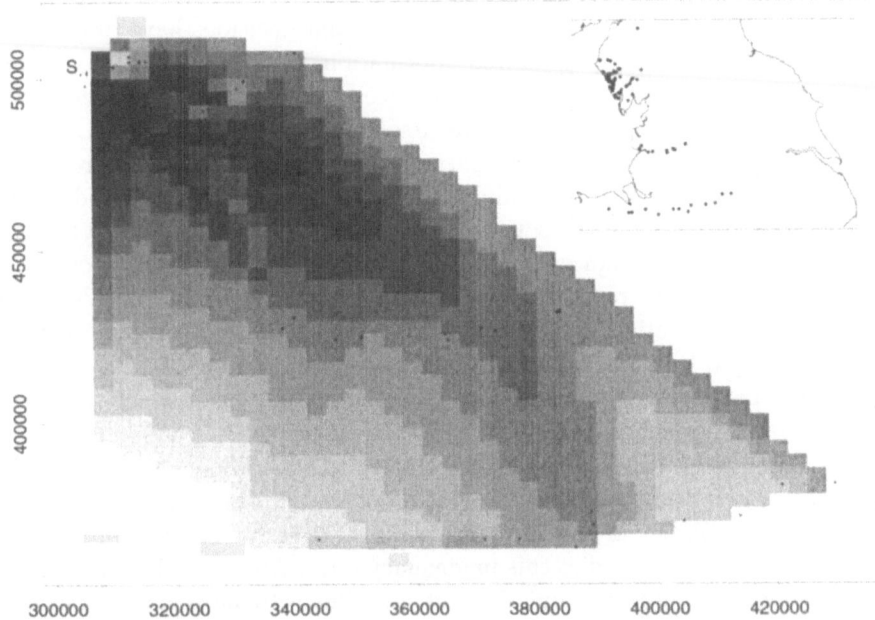

Fig. 11.1. Windscale data. Source marked as 'S'. Inset shows data area in the North-West of England.

inputs x represent the position of a point relative to the source of the release and the wind direction of the plume. As shown in Figure 11.1, whose scales are marked in metres, our prediction range for x is a sector that extends to approximately 150km south from the source and to more than 100km east. The calibration inputs t for the Gaussian plume model comprise the logarithm of the source term (total amount released) and the logarithm of the deposition velocity (rate at which material at ground level is deposited), and their true values for the Windscale incident comprise the calibration parameter vector θ. Other features such as wind speed, height of release and atmospheric stability could additionally be chosen for calibration but in this case are assumed to be known. The source term and deposition velocity are chosen for θ because the output is sensitive to the values of these parameters, and both are likely to be associated with significant uncertainty. It can also be argued that they are more linked to the particular nature of the event than other features of the release, e.g. the deposition velocity will vary for many reasons including the distribution of particle sizes released. For our single plume approximation of the Windscale data, we chose the wind direction which is believed to have accounted for the most significant release.

The output $\eta(x, t)$ from the code is the plume model's prediction of the logarithm of the level of deposition of at location x when the unknown parameters are t.

The simplicity of the Gaussian plume model makes it useful in the early stages of an accident assessment, when important decisions have to be taken quickly on the basis of very limited information. Compared with more elaborate dispersion models, very few inputs need to be specified in order to run the model. The Weerie model, a precursor of the dispersion model used in this study, successfully analysed part of the Windscale release, producing good agreement using a single plume along a relatively narrow band out to a distance of 50km from the source (Clarke 1974). Over the full area that we are attempting to fit, however, it is unlikely that such a simple model will provide an adequate representation of the overall deposition pattern. The wind changed direction during the release, the terrain ranged from coastal plain to rugged hills (so that the model's assumption of a single deposition velocity is particularly inappropriate) and the plume model is not valid over large distances or complex terrain.

When using such a simple code it is important to acknowledge that the model inadequacy term may be large. Any attempt to calibrate such a code, even with a large number of real measurements, will result in poor predictions unless we model and predict this inadequacy term. Another advantage of the plume code is that it is very quick to run, allowing us to make many thousands of runs very quickly, and we therefore assume it is a known function.

The observational data comprise $(x_1, z_1), \ldots, (x_n, z_n)$, where x_i is the variable input of the ith measurement and z_i is the corresponding actual ^{131}I log-deposition measurement.

Figure 11.2 shows the output from the plume model using input parameters of 1×10^{15} for the source term (total release of ^{131}I in Becquerels) and 0.004 ms^{-1} for the deposition velocity (i.e. 34.5 for the log source term and -5.52 for the log deposition velocity). These estimates are from published sources (Chamberlain 1981, 1991). Comparison with Figure 11.1 shows that the model seems to predict a plume that is too tightly focused.

11.3.3   Bayesian Analysis and Model Checking

We used the same figures of 34.5 and -5.52 as prior means for the calibration parameters. These are of course not strictly based only on prior information, but instead on analyses that others have conducted on similar data to ours. However, we assumed large prior variances of 5.0, as would be appropriate in the early stages of an accident, so that the actual estimates will have had little influence on the analysis.

Although we have a reasonably large number of observations of deposition, our objective is to be able to provide useful predictions of deposition based on as few observations as possible, so that the technique will be useful to the NRPB in the very early stages of a nuclear accident. Bayesian analyses were therefore carried out using the first 20, 30, 40, and 50 measured data points. Before presenting the principal results of this analysis, it is important to check the validity of our modeling assumptions. Since we have, in each analysis, not

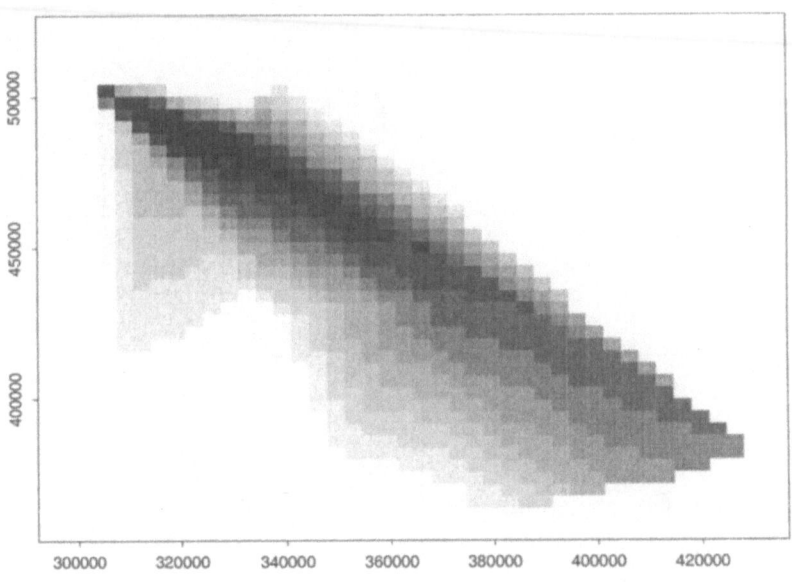

Fig. 11.2. Gaussian plume code output with fixed inputs.

used all the available data, this could be done by comparing the held-back data with their posterior predictive distributions, as was done in Kennedy and O'Hagan (2001). However, in practice this luxury will not be available to us, and we illustrate here a method based on cross-validation. With n data points, we computed the posterior predictive distribution of the i-th observation z_i by applying our method to the $n-1$ observations in which this i-th observation is omitted. Letting $\hat{z}_{-i}$ denote the mean of this predictive distribution, Figure 11.3 plots z_i against $\hat{z}_{-i}$ for each of the four sample sizes. This shows that the predictions from our model are in general agreement with the observed data, without any indication of systematic bias. However, the crucial question is whether the predictive errors shown in Figure 11.3 are of the order of magnitude indicated by the variances of the posterior predictive distributions. Letting the posterior standard deviation of the predictive distribution of z_i be s_{-i}, we computed standardised errors $e_i = (z_i - \hat{z}_{-i})/s_{-i}$. Figure 11.4 shows quantile-quantile plots of these standardised errors for each sample size. In each case the points lie close to the line of slope 1, showing that these errors are approximately $N(0,1)$ distributed. We believe that such plots support clearly the validity of our underlying model.

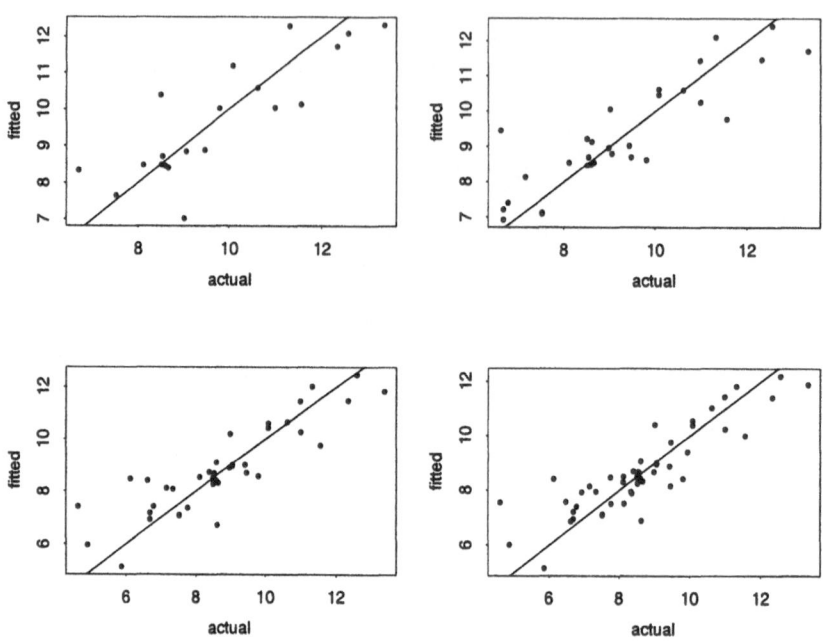

Fig. 11.3. Cross validation predicted vs. actual depositions based on $n = 20$, 30, 40, and 50 points.

11.3.4   Results

The same cross validation approach can be used to compare the accuracy of our method with other strategies. Prediction accuracy for each strategy was assessed by using leave-one-out cross validation and calculating Root Mean-Square Error. With n data points used, this is

$$RMSE = \left[n^{-1} \sum_{i=1}^{n} (z_i - \hat{z}_{-i})^2 \right]^{1/2} \qquad (11.1)$$

where z_i is the true log-deposition at variable input x_i and $\hat{z}_{-i}$ is the estimate from the strategy in question using all the data except point i. The following prediction strategies were compared.

- Strategy 1. Using the code predictions alone, with fixed θ.
- Strategy 2. Using model inadequacy corrections, with fixed θ.
- Strategy 3. Using model inadequacy corrections and Bayesian calibration for θ.

Strategy 3 is therefore the fully Bayesian analysis, whose validity was confirmed in the preceding subsection, while Strategy 1 gives the predictions

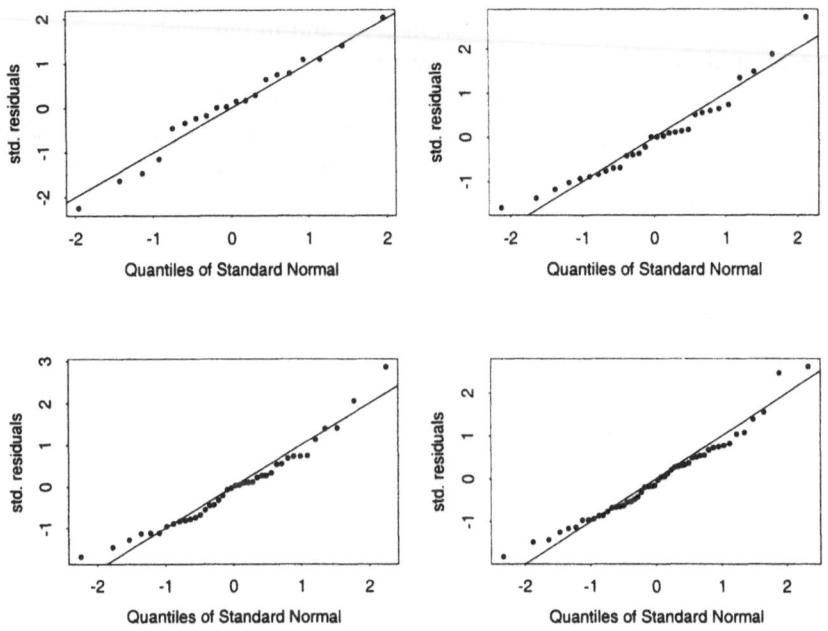

Fig. 11.4. Cross validation Quantile-Quantile plots based on $n = 20, 30, 40$, and 50 points.

shown in Figure 11.2. Strategy 2 is equivalent to the fully Bayesian analysis but with a very strong prior distribution centred on the prior estimates of log source term and log deposition velocity. It therefore does not use the calibration part of our method, but does use the model inadequacy part.

Table 11.1. Root Mean Squared Error of prediction for the Windscale samples

Strategy	1	2	3
$n = 20$	2.98	1.93	0.94
$n = 30$	3.09	1.79	0.87
$n = 40$	2.85	1.80	0.96
$n = 50$	2.83	1.95	0.88

Table 11.1 shows the RMSE values for each strategy and each sample size. As expected, the code output with inputs fixed at their prior means produced the largest RMSEs. Allowing model inadequacy significantly reduces these errors, primarily by 'correcting' the code close to the source. More importantly, Strategy 3 gives a very noticeable further reduction in RMSE values. The reason for this is clearer if we examine the separate effects of calibration and model inadequacy in the prediction of $z(\cdot)$.

Figure 11.5 shows the calibrated code based on the 50 point dataset, without the inadequacy term. The posterior distribution of θ gives most probability to small values of the calibration parameters, and in particular the posterior distribution of the log source term is concentrated on much lower values than the prior estimate of 34.5 taken from the literature on the Windscale accident. This has had the effect of minimising the code's influence in the predictions. The observed log-deposition data is then almost entirely explained by the model inadequacy term.

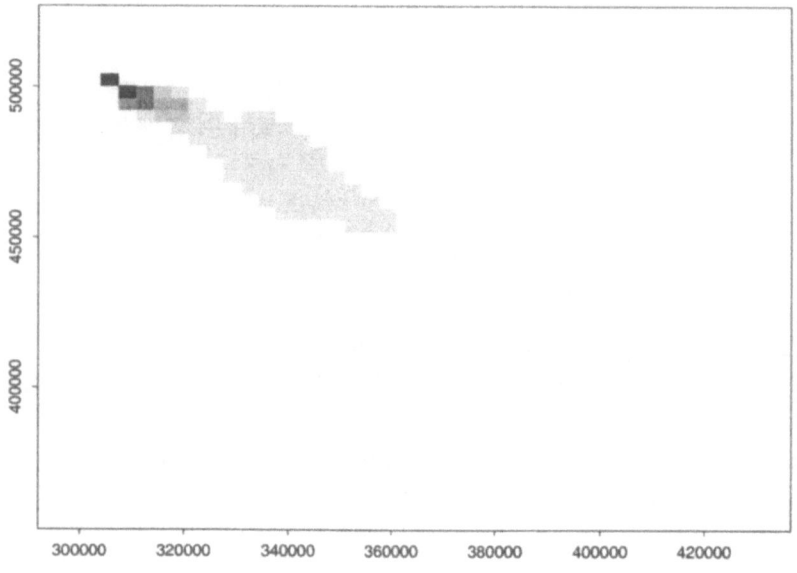

Fig. 11.5. Calibrated code output, without model inadequacy correction.

This example illustrates well the interplay between the computer code $\eta(\cdot,\cdot)$ and the model inadequacy $\delta(\cdot)$ in our model. The meteorological conditions during the Windscale accident made it unlikely that a simple Gaussian plume model would predict the true radioactive deposition accurately. Our analysis confirms this in the dramatic way illustrated by Figure 11.3. The computer code fails to fit the data, and the freedom allowed to the model inadequacy term in our model has been exploited fully to fit the data primarily as smoothed residuals from what little is explained by the calibrated code.

11.4   Discussion

Some important general issues are raised by this example.

The calibration estimate of the log source term is very small. However, this does not imply that an estimate of the total amount of radioactivity released in the Windscale accident would be dramatically different from previous estimates. As we have seen, the calibrated code only accounts for a small part of the estimate of the true log deposition $\zeta(\cdot)$, and the magnitude of the model inadequacy term implies a further very substantial emission. The process of calibration in general finds the values of calibration parameters that best fit the observed data in some sense, and these values need not be the 'true' values of the input variables as understood by the developers of the computer model. This distinction becomes even more important with our addition of the model inadequacy term $\delta(\cdot)$, and has been discussed in Kennedy and O'Hagan (2001).

Our analysis has been developed for the purpose of making predictions of the real-world phenomenon in the same context as the observed data (i.e. with the same values of the calibration parameters), at unobserved values of the variable inputs. For the Windscale example, this means predicting deposition from the Windscale accident at unobserved sites. We believe that the analysis is shown to be effective in this predictive role, but great care should be taken over interpreting the posterior estimates of calibration parameters. The source term has a direct physical interpretation as the total amount of radioactivity released in the form of the ^{131}I radionuclide, and provides a direct measure of the scale of the event. However, this is not the same as the fitted value of the 'source term' parameter in the simple dispersion model. Both are potentially useful and ideally will be numerically quite close, but they are not synonymous. By introducing an inadequacy term the 'source term' that scales the simple model loses its numerical similarity to the real source term. Information about the real source term would have been valuable, and failure to gain this experience may be seen as a disadvantage of our method (although some of that information may potentially be regained from the model inadequacy term). However, it is clearly unrealistic to hope to learn about the true values of these physical variables from a computer model that only imperfectly represents the real phenomenon.

This example does not tell us anything new about the Windscale accident. Work is in progress to carry out a more extensive analysis of these data. Results of our initial analysis are presented here primarily to illustrate the behaviour of Bayesian calibration when we try to calibrate a computer code that does not fit the observed data.

Work is also progressing on applying these methods to other kinds of computer model.

References

Arnold, L. (1992). *Windscale 1957: Anatomy of a Nuclear Accident.* Macmillan.

Bates, R. A., Buck, R. J., Riccomagno, E. & Wynn, H. P. (1995). Experimental design and observation for large systems. *J. Roy. Statist. Soc. B*, **58**, 77–94.

Chamberlain, A. C. (1981). *Emission of Fission Products and Other Activities During the Accident to Windscale Pile No. 1 in October 1957.* Atomic Energy Research Establishment Memorandum 3194.

Chamberlain, A. C. (1991). In *Proceedings of Comparative Assessment of the Environmental Impact of Radionuclides Released During Three Major Nuclear Accidents: Kyshtym, Windscale, Chernobyl,* Volume 1, 303–308. Commission for the European Community, EUR 13574.

Clarke, R. H. (1974). An analysis of the 1957 Windscale accident using the WEERIE code. *Annals of Nuclear Science and Engineering,* **1**, 73–82.

Clarke, R. H. (1979). *The First Report of a Working Group on Atmospheric Dispersion: A Model for Short and Medium Range Dispersion of Radionuclides Released to the Atmosphere.* London: Her Majesty's Stationery Office.

Craig, P. S., Goldstein, M., Rougier, J. C. & Seheult, A. H. (2001). Bayesian forecasting using large computer models. *J. Amer. Statist. Assoc.* (in press).

Craig, P. S., Goldstein, M., Seheult, A. H. & Smith, J. A. (1996). Bayes linear strategies for matching hydrocarbon reservoir history. In *Bayesian Statistics 5*; Bernardo, J. M., Berger, J. O., Dawid, A. P.& Smith, A. F. M. (eds.). Oxford: University Press. 69–95.

Currin, C., Mitchell, T., Morris, M. & Ylvisaker, D. (1991). Bayesian prediction of deterministic functions, with applications to the design and analysis of computer experiments, *J. Amer. Statist. Assoc.,* **86**, 953–963.

Haylock, R. and O'Hagan, A. (1996). On inference for outputs of computationally expensive algorithms with uncertainty on the inputs. In *Bayesian Statistics 5*; Bernardo, J. M., Berger, J. O., Dawid, A. P.& Smith, A. F. M. (eds.). Oxford: University Press. 629–637.

Kennedy, M. C. and O'Hagan, A. (2001). Bayesian calibration of computer models (with discussion). *J. Roy. Statist. Soc. B,* **63**, 425–464.

Morris, M. D., Mitchell, T. J. & Ylvisaker, D. (1993). Bayesian design and analysis of computer experiments: use of derivatives in surface prediction, *Technometrics,* **35**, 243–255.

NRPB (1990a). Principles for the protection of the public and workers in the event of accidental releases of radioactive materials into the environment and other radiological emergencies. *Documents of NRPB,* **1**, No.4, 1–4.

NRPB (1990b). Emergency reference levels of dose for early countermeasures to protect the public. *Documents of NRPB,* **1**, No.4, 5–33.

Oakley, J. E. and O'Hagan, A. (2000). Bayesian inference for the uncertainty distribution. Tech. Rep. University of Nottingham, Statistics Section.

O'Hagan, A., Kennedy, M. C. & Oakley, J. E. (1999). Uncertainty analysis and other inference tools for complex computer codes (with discussion). In: *Bayesian Statistics 6*; Bernardo, J. M., Berger, J. O., Dawid, A. P. & Smith, A.F.M. eds., 503–524. Oxford: University Press.

Sacks, J., Welch, W.J., Mitchell, T.J. & Wynn, H.P. (1989). Design and analysis of computer experiments, *Statist. Sci.*, **4**, 409–435.

Saltelli, A., Chan, K. & Scott, E. M. (eds.) (2000). *Sensitivity Analysis*. New York: Wiley.

Stein, M. L. (1999). *Interpolation of Spatial Data: Some Theory for Kriging*. New York: Springer-Verlag.

12   The Realities of Decision Making on Risks

Jim McQuaid

Department of Mechanical Engineering, University of Sheffield,
Mappin Street, Sheffield S1 3JD, UK. **

Abstract. The interaction of man with natural and manmade hazards is subject
to many uncertainties. Risk assessment in its general sense is the study of that
interaction in order to inform decisions about what should be done in any given
context. Advances in risk estimation methodologies have been paralleled by equally
important changes in society's response to risk issues. For a long time, risk was
seen as no more than an objective measure of the possibility of identified physical
harm and exclusively a matter for expert assessors. Any reluctance to accept the
expert view was dismissed as irrational. But several factors have combined to make
decision making on risk a fertile and complex subject of debate. These include:

- growing understanding of the influences that colour people's perception of risks;
- heightened awareness of the dependency of expert assessments on judgement;
- consequent demands for greater openness and inclusion of stakeholder values
  in framing the issues and arriving at a balanced decision, and
- incessant media clamour driven by the difficult politics of the equitable distribution of costs and benefits.

In threading a way through this minefield, the decision maker can succeed only with
expert advice that is robust and tractable to public scrutiny. Various developments
relevant to that aim will be discussed.

12.1   Introduction

In addressing the subject of decision making on risks, I am drawing upon two
perspectives gained from my past experiences rather than from my present
affiliation. It would be useful to say something about these so that my cre-
dentials, such as they are, will be known, though the views and opinions I
express in this paper are entirely my own.

The first perspective comes from having been Chief Scientist and Director of
Science and Technology of the Health and Safety Executive of Great Britain.
The HSE is the government regulator of all industrial activities other than
the marine and aviation sectors. Its remit covers risks to the public who may
be affected by those activities in addition to those who are occupationally
exposed to the risks. The remit is reflected in HSE's mission which is 'to ensure that risks to people's health and safety from work activities are properly
controlled'.

** formerly with the Health & Safety Executive of Great Britain.

This very wide remit derives from powers under the Health and Safety at Work Act (HSWA) of 1974. The HSWA established HSE as unified regulator (except in the areas noted above), replacing a spread of regulatory responsibilities across different government departments. In addition, the HSWA itself changed the philosophical basis of state regulation in line with the recommendations of the report of the Robens Committee on Health and Safety at Work (Robens Committee 1972). The change of philosophy introduced a new approach to making decisions on how the state should intervene to achieve the desirable social and business objective of improving health and safety performance. Further discussion of the change and the reasons for it is appropriate for this paper and will be given in the section that follows.

The second perspective that I bring comes from having been chairman of the Interdepartmental Liaison Group on Risk Assessment (ILGRA). This body is comprised of representatives of all UK government departments involved in the development of policies and practices on risk issues (McQuaid & Le Guen 1998). It exists to promote coherence and consistency in the process of decision making across government on matters pertaining to risks. It analyses and deliberates on issues of common interest, promotes research on behalf of government departments and publishes guidance. It prepares reports for Ministers which are published (ILGRA 1996, 1998) and obtains Ministerial endorsement for its programme, though it relies on informal mechanisms for gaining the cooperation and collaboration of participating departments acting in their individual and collective interests. ILGRA provides an important means for ensuring that the changing nature of risk-based decisions - the main theme of this paper - is properly considered and reflected in the approaches of government departments. An important outcome of the work of ILGRA was a commitment in the Modernising Government Action Plan (Cabinet Office 2000) that all departments should publish statements describing their decision making frameworks and some have been published (for example, HSE 1999a; Department of Health 1999; MAFF 2000). A further commitment is that government will publish a Statement on Risk in 2001, as presaged in MAFF(2001). Overall, it may be said that there has been a significant move towards openness and public participation on risk issues in the UK in recent years. This move responds to the realisation that controversies surrounding the uncertainties and value-laden nature of risk issues cannot be damped down by objective or pseudo-objective arguments but rather that such arguments must be complemented by greater openness and participation, about which more will be said later.

12.2   Decision Making on Health and Safety

The framework for government decision making on risks to health and safety from work activities in Great Britain rests on the affirmation by the 1972 report of the Robens Committee that the primary responsibility for control should lie with those who create the risks and those who work with them. This was in contradistinction to the view that prevailed throughout the course of development of industrial health and safety legislation over the preceding 150 years. That view was, whether made explicit or not, that the responsibility rested with the state. The state would prescribe what had to be done in every detail, building upon the lessons from actual experience of harm and from the evolution of methods to prevent the harm. The state would back up its prescriptions by inspection of workplaces to check compliance. There were several consequences. First, there was an ever increasing body of legislation much of which rapidly became obsolete with the advance of technology and changing industrial structures and practices. By around 1970, the realisation had sunk in that the complexity and elaborate detail of the law was too much to bear and was proving ineffective. The second was that there was widespread apathy in industry about health and safety performance, inhibiting improvement through innovation and anticipation. The third was that inspection by the state for compliance was proving unequal to the task, with visits by the inspectorates to the generality of workplaces being infrequent, or in some instances never actually happening. Regulatory decision making, however, was straightforward as mere drafting of prescriptive standards. Compliance with the standards would satisfy the law. No harm would occur if nothing ever failed and people behaved as required and never made mistakes.

The placing of responsibility on identified duty holders (the 'regulated') was to be achieved by appropriate framing of the legislative regime, by a changed relationship between the regulator and the regulated and by new levers in the regulator's armoury of incentives, all designed to give a greater emphasis to prevention of harmful occurrences rather than retribution after the event. There were concomitant changes to the way decisions were made about regulatory requirements and to the capability of the regulator to engage more constructively with the risk creators in industry and those affected.

Thus, in summary, the new regime :

- framed the legal requirements in terms of general duties unrelated to hazard or risk, backed up by regulations which set goals to be achieved, with a reserved capacity to prescribe (or sometimes to proscribe) in the case of incontrovertible hazards or when politics demanded a step change in industry behaviour;
- gave considerable discretion to the regulator and to the regulated on approaches to the control of risks to meet the legal requirement that they should be reduced so far as is reasonably practicable. The result was a focus on dialogue, initially between the regulator and the regulated but more recently subject to pressure to be more inclusive of all appropriate interests in the pursuit of openness about decision making;
- introduced variety into the process of enforcement of the law, ranging from the serving of prohibition notices where there was an immediate threat to health and safety, through improvement notices requiring action

to be taken in a specified time, to the provision of guidance and advice on standards of good practice, supported by an extensive programme of publications drawn up in consultation with industry. Prosecution in the courts was of course retained for severe or persistent offenders, since breaches of health and safety requirements attract criminal sanctions;

- emphasised the importance of methodological assessment of hazards and risks and, where established and authoritative good practice did not exist, the adoption of systematic methods of balancing the achievement of risk reduction with the costs of doing so;
- introduced an extensive system of advisory committees to draw both sides of industry into deliberation with the regulator on the development of methods for assessment of risks and cost-effective systems for their control;
- expanded the capability of the regulator on all aspects of science and technology pertaining to risks, from research on causation and mitigation to proactive development of safe and healthy systems, all in order to inform the framing of legislation, to develop guidance for publication and to equip the regulator to engage constructively in the dialogue with the regulated. There was a considerable expansion in the scientific and engineering investigation of accidents, reflecting the needs of public inquiries, the benefits of wide promulgation of lessons learned and the identification of areas where research was needed to stimulate improvement in performance.

All of these changes were progressively implemented and had implications for the process of decision making. A landmark in that evolution was the publication by HSE in 1999 of a detailed exposition on how it makes regulatory decisions and the many considerations taken into account. The document was published for public discussion (HSE 1999a) and was in effect a rendering to public account of its stewardship. The document was a response to the enhanced demand for explanation of the basis of decision making on risks since the enactment of HSWA and described the influences that now prevail – though it should be said that much of what is regarded as correct nowadays was identified by the Robens Committee and further described by HSE's first Director General (Locke 1981).

12.3   Changing Nature of Risk Debates

It is a matter of common observation that risk issues nowadays occupy the attention of the public and the media to a seemingly inordinate extent. This is evidenced by the way risk issues frequently capture and retain headline attention. Government and industry decision makers are constantly assailed with demands for explanation which, in the terms in which they are able to offer it, seem to do little to assuage public concerns.

The strong public reaction to risk issues is sometimes presented as a modern phenomenon though history suggests otherwise. The development of the flame safety lamp for coal mines in the early 19th century resulted directly from representations in 1815 to Sir Humphry Davy by a society of lay people concerned about the toll of deaths in explosions (McQuaid 1997). The public outcry that followed a railway disaster at Armagh in 1889 forced Parliament to enact draconian legislation specifying changes to working practices, much resisted by the railway companies, within two months of the disaster. In 1890, the Home Secretary, following questions in Parliament, took the extraordinary step of inviting suggestions from the public about what should be done to reduce the incidence of great coal mine explosions. This led directly to legislation requiring action to control the hazard of coal dust which was known to contribute to the propagation of explosions (McQuaid 1997).

However, the risk debates of the present day are often motivated by fears about perceived threats as distinct from concerns about actual evidence of harm, though of course exceptions arise as a result of particular accidents such as recent occurrences on the railways. There is often said to be a loss of trust by the public in the institutions of the state, though the examples quoted above show evidence of a distinct lack of trust in past times. In the face of actual experience of harm, it would seem that not much has changed. The public, media and concerned groups can be mobilised to bring pressure for change, though the power of modern communications can increase the intensity of the pressure and thereby help to achieve change more quickly to deal with the incontrovertible evidence – though there is no recent instance of legislation being enacted within two months as happened after the Armagh disaster. Where it is a matter of an anticipation of harm, the debate is usually wide open. There is certainly an unwillingness by the public to place reliance on the delivery of corporate responsibility by the risk creators. This extends to reluctance to countenance the unfettered influence of the risk creators on the government's scientific advisory machinery.

The arguments about harms that are anticipated or hypothesised though not yet experienced are of a different character to those surrounding the causation of accidents and ill-health. The role of science and technology is on relatively sure ground in the latter circumstance given the availability of evidence and the scope for closely defined investigations to resolve differences between experts and to develop defences to prevent a recurrence. But in the former circumstance, the inevitable uncertainties, the difficulty or impossibility of acquiring evidence and hence the need to exercise expert judgement elevate the problems into the realm of 'trans-science' as coined by Weinberg (1972) or 'post-normal science' by Funtowitz and Ravetz (1990). The nature of the problem means that predictions by individual experts of events that may be conceived as possible will differ, sometimes substantially, in terms of their scale and likelihood.

The influence of science and technology in debates about what should be done in the case of such issues is increasingly questioned, leading to contention as to the basis for regulatory decisions on the measures to be taken. The public do not see it as legitimate that a purely scientific view should dominate the decision. Even less so is that the scientific view should determine the decision as a result of pressure on scientific advisers by decision makers to resolve a policy dilemma on what should be done, with the scientific advice presented as objective, without any qualification about the uncertainties. An even greater difficulty for decision makers arises when fears or suspicions are expressed by single issue pressure groups and there is little or no scientific basis for concluding either way, other than on a balance of probabilities. This balance may be based on subjective interpretation of such information as is available or extrapolation from experience in related fields or, in the extreme, nothing more is possible than a visceral feel for what might or might not happen. Pressure then arises for the strict application of the Precautionary Principle (Gibson & Kass 2001) as a way of achieving certainty by an outright ban on the causative agent. The proper influence of science and technology in these various facets of risk debates will be further discussed in more detail below.

Thus the most acute contention that increasingly arises, and not helped by the uncertainties of the science, focuses on what Locke has called the 'politics' which he defined as 'the formal institutions and procedures established by society to deal with differences of interest between various groups in that society' (Locke, 1981). The real political battleground is where there is a reaction to the imposition of risks, with a lack of confidence in the robustness of the procedures by which the public interest – or more accurately, the diversity of public interests rather than an average 'public interest' decided by the nanny state – is given due weight in the policy deliberation. Interest groups can be very effective in promoting their own value systems in the public arena and these may have more resonance with the public than those applied by the decision makers. The latter values will be embedded in the protocols for the valuation of the benefits of reduction of risk and the costs of achieving that reduction and, more fundamentally, in the important policy issue of whether the concern of the public should be represented or framed by posing the question 'what is the risk?' The disposal of the Brent Spar oil storage platform is the leading recent example. Furthermore, the extent to which the vulnerable, the disadvantaged and the uncompensated victims do not experience justice and fair play by comparison with the beneficiaries of a risk management policy is a powerful mobiliser of public controversies. On the other side of the debate are those concerned that undue amplification of the downsides of a risky undertaking is a constraint on economic and social progress.

Questions that inevitably arise in debates on risk issues are:

- who influences the way the issue is framed? Is it influenced by bias based on preconceived ideas about what would be a workable solution in the sense, for example, of being easily enforceable and not burdensome on powerful vested interests?
- who sets the priorities and on what basis? Is too much weight given to supposedly objective analysis and mechanistic criteria of what society should be willing to tolerate?
- whose views count in the making of the decision? Are the procedures rigorous enough in controlling the potential for domination by vested interests of whatever hue?

Aspects of these questions will be considered in one way or another in what follows.

12.4 Changing Influence of Science

Popular belief has traditionally conferred considerable trust in science as giving the right answers, a view fostered by the type of science taught in the educational curriculum. This trust has been bolstered in recent decades by experience of the demonstrable successes of science. Decision makers naturally sought to take advantage of this trust in communicating to the public about risk issues. Such was the position that prevailed up to the 1970s and still prevails to some extent. But the approach came increasingly to be questioned, particularly with the reluctance of the public to share the experts' views on safety in critical areas of policy such as nuclear power. This led to much activity on what constituted an 'acceptable' risk and an early attempt to understand the public's view by examination of the role of voluntariness on acceptability of risks (Starr 1969). This was paralleled by efforts to promote the message to scientists that it was futile and irresponsible to cloak assessments of risk by scientists as an exercise that conformed to the rigours of the scientific method. Notable amongst these efforts were the publications of Weinberg (1972) in the area of nuclear power and by the Council for Science and Society (1977) motivated by the then developing science of genetic manipulation. The main message was that the practical impossibility of 'proving' the predictions of the risk assessment of a large scale hazard e.g. of the probability of catastrophic failure of a pressurised water reactor, demanded that scientists should expose the uncertainties to public scrutiny and learn to accommodate to the need to engage with the public. As Weinberg (1972) stated: 'We in the technological and scientific community value our republic and its workings. But when what we do transcends science and when it impinges on the public, we have no choice but to welcome the public - even encourage the public - to participate in the debate. Scientists have no monopoly on wisdom where this kind of trans-science is involved: they will have to accommodate to the will of the public and its representatives. The republic of trans-science, bordering as it does on both the political republic and the republic of science, can be neither as pure as the latter nor as undisciplined as the former. The most science can do is to inject some intellectual discipline into the republic of trans-science; politics in an open society will surely keep it democratic.'

It was only much later, at the end of the 1990s, that exhortations such as Weinberg's came back into prominence following a period in which quantitative risk assessment flourished as the apparent answer to the decision maker's prayer. This was influenced by the development in the interim of knowledge and hypotheses concerning public reaction to risk issues and the repeated evidence of failure to convince the public by communication designed to improve the understanding of the public on the complexities of risk issues. A particularly cogent appraisal of the reasons for that failure was provided by Waldegrave (1987) from the standpoint of an experienced politician, echoing the earlier arguments of Locke as the experienced policy maker and regulator (Locke 1981).

12.5   Public Responses to Risk Issues

Enlightenment on the reasons for the increasing inability of scientific arguments, represented as objective truths, to persuade the public was provided by research on the social and psychological factors that influence the public perception of risk. The psychometric paradigm was developed (Fischhoff *et al.* 1978) as a mapping of the characteristics of risks and of the context in which they are experienced as they affect the public's perceptions and hence views on how rigorously the risks should be regulated. These characteristics were determined by research in which members of the public were invited to express their preferences across a variety of risks. They were found to include the extent to which the risks are known and understood, how well individuals can control (or believe that they can control) their exposure to the risks, the feelings of 'dread' aroused by the risks, as well as the voluntary character of the risks as earlier examined by Starr (1969).

Cultural theory, with its roots in social anthropology, was also developed (Douglas & Wildavsky 1982) to explain how different types of rationality in human behaviour are influential in the acceptance or otherwise of risks or of structures of governance. The underlying precept is that cognitive science, from which the psychometric methodology was developed, is based on the proposition that the risk receiving public is an aggregation of unconnected individuals (Douglas 1994). Each is assumed to be judging risks on their own whereas they are social beings and consult about risks. Cultural theory asserts that the reactions of individuals to technological risks are determined both by the values they hold and by the model of society to which they subscribe. The two cannot be separated; the latter determines the former. Cultural theory elaborates on the types of viable society and the risk taking behaviour of the members of each type. For further details, see Douglas (1994) and Adams (1995).

The influence of the media and in particular the interaction between public perceptions and media coverage were examined in the context of the social amplification of risk hypothesis (Kasperson et al. 1988). This hypothesis attempts to explain why experience of different risks can generate very different intensities of public debate. These are not proportionate to any numerical estimate of the actual risks but rather are hypothesized to depend on the terms and graphic imagery in media messages, the way these messages impact on the public and on reinforcement by the feedback mechanism provided by media call lines and on-the-spot interviews. A comprehensive programme of research for UK government departments on the extent to which risk debates in the UK can be explained by the hypothesis has recently been completed. One of the main findings of the research is a refutation of any suggestion that lay publics are passive recipients of expert risk knowledge (Petts et al. 2001). All of the above developments provide useful explanations of public responses for the purpose of risk communication and have been reflected in one way or another in guidelines produced for government policy makers. However, their predictive capacity for use in decision making has yet to be demonstrated.

12.6   Features of the Public Debate on Risk Decision Making

The evolution of the structure of risk debates has reflected the different interpretations of risk issues by scientists/engineers and by the public/media as described above. The divergence of views continues to bedevil and confuse the public debate and to inhibit the search for workable and sustainable outcomes in terms of what should be done. Despite Weinberg's exhortation in 1972, the debate continued to be conducted on separate lines throughout the 1980s and much of the 1990s. A publication by the Royal Society in 1992 (Royal Society 1992) exemplified the separation, with papers on the opposing views and with no meeting of minds.

Thus the public debate has until recently been largely conducted from two standpoints reflecting different views of how decisions on risk issues should be made. A technocratic standpoint argued that the best resources of science and engineering are capable of estimating the probabilities of events and the scale of possible damage, of deriving monetary measures of what the public are prepared to pay to reduce risks and hence of arriving at a decision rule on what is 'acceptable'. There are refinements on this basic methodological structure but the main message was that decisions should be based on hard 'rationality'. Anything else was portrayed as a distortion of the economic efficiency of allocation of resources, to the ultimate detriment of society.

The other standpoint argued that risk assessment does not lead to objectively 'right' answers, especially where there is great uncertainty, and that decisions on risk issues involve both practical and moral judgements. These include:

- what things matter to society;
- what arouses fear and how should this be taken into account;
- who is to be believed in arguments where protagonists come to different conclusions on the basis of the same information;
- what is fair, ethical and equitable;
- who benefits from the decision and, perhaps most importantly,
- how society wants to be governed and who decides the priorities.

This mistrust of experts has a long tradition. Lord Salisbury evocatively described it thus in the 1880s:

'No lesson seems to be so deeply inculcated by the experience of life as that you never should trust experts. If you believe the doctors, nothing is wholesome; if you believe the theologians, nothing is innocent; if you believe the soldiers, nothing is safe. They all require to have their strong wine diluted by a very large admixture of insipid common sense.'

Harold Laski entered the fray in 1932 with a critique of experts pronouncing on matters of social policy (Laski 1932). He prophetically described the problem as he saw it thus:

'The experience of the expert is so different, his approach to life so dissimilar, that expert and plain man are often impatient of each other's values. Until we can somehow harmonize them, our feet will be near to the abyss.'

It is hardly to be expected that the public will understand complex scientific concepts but they are not necessarily as ignorant or irrational as the experts might want to argue in the context of the public's judgements about risk issues. The experiences of recent years have demonstrated that people display great interest and wish to be involved in the non-scientific issues raised by the questions listed above. A particular example is the perception of non-independence of science and government from commercial interests and the need for assurance that the information and advice on which decisions are made on their behalf take proper account of the diversity of public views. This assurance formerly rested on trust but it is now widely accepted that it must be provided explicitly by evidence of competence, independence, consistency and the like.

Although there are still protagonists firmly wedded to one view or the other, there has more recently been a proliferation of official reports promoting a middle-ground position, much on the lines advocated by Weinberg. These are exemplified by high-level publications of the US National Research Council (Stern & Fineberg 1996) and the US Environmental Protection Agency's Presidential Commission (EPA 1997). A report from a group led by the UK Treasury surveyed the economic, psychological, ethical and administrative aspects of safety standards (Her Majesty's Treasury 1996). The report concluded that a fully 'rule-based' approach to safety regulation, where all regulations are set according to universal formulae quantifying and valuing costs and benefits, would be unrealistic. Nonetheless common frameworks could and should be developed for all safety regulation, despite the wide differences in available data and in attitudes towards different kinds of risks, so providing a common basis for policy judgements and standard setting. The various reports from ILGRA and UK government departments and agencies cited earlier reflected a growing appreciation across government of the need for public engagement in the application of such a framework and transparency of the procedures for the delivery of the engagement and application. A similar appreciation is reflected in a report in the Netherlands (Health Council of the Netherlands 1996). At the international level, a European Concerted Action supported by the European Commission addressed the social management of risk and developed a framework for risk governance, reported in TRUSTNET 1999. The report noted that, despite the wide range of activities and risks covered in the programme (health, technological risks, natural hazards, nuclear risk, etc), a common analysis emerged though it was also noted that significant cultural and semantic differences sometimes remain between the approaches in the various European countries.

In the UK and other countries, as well as international institutions, there has also been a raft of related publications on the subject of the nature of scientific advice and the procedures for its provision to government (Royal Society 1997, 1999; OXERA 2000; Royal Commission on Environmental Pollution 2000; House of Lords 2000; MAFF 2001; OST 1997). The relevance of these publications is that the scientific advice is usually being sought by government on the interpretation and characterisation of risk issues as an input to policy decisions on the management of the risks. The general conclusion, exemplified by House of Lords (2000), was that direct dialogue with the public should move from being an optional add-on to science-based policy making. Any public dialogue should be conducted in good faith and its aims and in particular its role in the policy process should be clear from the start. The middle ground position advocated in one way or another by the above collection of papers and reports recognises the limitations of risk assessment but also its strengths when properly conducted. These limitations and strengths will be further discussed below. The advocates of the middle ground would also seek engagement with and education of the public. Information about the public's perception of risk would form useful and necessary inputs, for example in setting tolerability and risk aversion criteria appropriate to different risk situations. It would be a necessary condition that these inputs would involve direct participation by the public. This would be achieved through recently developed participative mechanisms such as consensus conferences and focus groups, in addition to the traditional processes of consultation and negotiation, and by lay representation on advisory committees.

12.7   Risk-Based Approach to Regulation

The evolution of the risk-based approach put in place in Great Britain by the HSWA has taken place in parallel with the various changes in roles, practices and understanding described above. The approach requires an anticipatory and creative attitude to thinking about the range of possible ways in which harm can occur, with the risk expressed as the combination of consequences and likelihood. The emphasis is on achieving reduction of risk through the setting of regulatory goals. This emphasis does not necessarily require a numerical estimate of the risk but rather a determination to strive to reduce the risk to the point where further effort is not considered to be reasonably practicable. This is the underlying philosophy of the UK regulatory approach (Bacon 1994). It places requirements on duty-holders (i.e. those with duties under the law) that are generally qualified in terms of ensuring health and safety 'so far as is reasonably practicable (SFAIRP)'. This is usually rendered as reducing risks 'as low as reasonably practicable (ALARP)'. Guidance exists from case law as to how the courts interpret the meaning of the reasonably practicable qualification. This guidance is reviewed in the HSE explanatory document referred to earlier (HSE 1999a), setting it in the context of HSE's approach to regulatory decision making.

Consideration of risk reduction towards the ALARP objective can lead to much debate about what is or is not reasonably practicable. The argument is not the equivalent of contending that something is safe or not safe but whether it is being made safe enough, using the language preferred in public debates. It is the function of the regulator to establish principles by which it may be judged whether indeed enough is being done, bearing in mind the decisions of the courts. The regulator must also have regard to the principles of procedural justice and good regulation, encapsulated as satisfying criteria of transparency, consistency, targeting, proportionality and accountability.

Considerable effort has been devoted to the development of an analytical procedure for assisting judgements on whether risks to an exposed individual are ALARP, where the nature of the individual's exposure must necessarily be defined as part of the procedure. The procedure is primarily based on maximising utility, i.e. the effectiveness of expenditure in relation to the reduction in risk that is achieved, though other criteria may also be included, such as satisfying the need for equity by setting a maximum to the risk to which an individual may be exposed. The procedure defines a risk that is ALARP as being 'tolerable' in the sense that it can be lived with for the time being in the knowledge that all that can reasonably be done to reduce it is being done and that it will continue to be kept under review (HSE 1992). Application of the procedure requires information or protocols for ascribing a monetary value to marginal reduction in risk. Although it is perhaps unfortunate in terms of public understanding, it is conventional to use a monetary value unitised to that for an individual life, called the Value of Statistical Life (VoSL). It is derived by aggregation of the values at the margin. The VoSL – frequently abbreviated incorrectly to 'Value of Life' – is central to the balancing of costs and benefits. Its numerical magnitude is obtained from empirical studies of the public's willingness to pay for defined safety improvements (Jones-Lee 1994). Most attention has been given to the value appropriate to the assessment of the benefits of road safety improvements. The value so derived has been taken as a benchmark to which multiplying factors may be applied to attempt to reflect the public's known aversion to other kinds of risk, as described earlier. However, there is little empirical evidence to support this practice and it has recently been challenged on the basis of research which appears to show that the public are not willing to differentiate between risks by paying more themselves for reducing the kinds of risk to which they are held to be averse, for example railway accidents (Chilton *et al.* 2000).

The risk to an individual has for long been the only consideration in the technical setting of safety standards for ensuring compliance with statutory requirements. Of increasing importance in recent years is the more difficult question of how to incorporate the undoubted societal dimension that pertains to some risks through the manner in which they can arouse extreme societal concerns, for example through their potential for causing multiple fatality accidents or irreversible effects on future generations. There are at present no statutory requirements explicitly addressing such concerns over and above their representation by the risk to the individual members of society. The setting of safety standards covering societal concerns must necessarily have a political or commercial motivation, driven by the need to minimise political or commercial fallout. There are no accepted general procedures for quantifying such societal concerns that can be reflected into risk assessment methodologies with the exception of multiple fatality accidents (Ball and Floyd 1998). In that case, the statistical profile of accident sizes for the type of accident in question is compared to the profiles exhibited by naturally occurring accidents of varying scale and frequency. The procedure is akin to comparing individual safety risks to the commonly occurring risks to individuals in their everyday lives, such as being struck by lightning or the benchmark risk of dying from all causes. The comparisons are intended to put the risks into perspective as an aid to the making of the decision on what, if anything, is to be done. There is a lack of convincing empirical evidence to support the quantification of arguments about societal concerns, due no doubt to the considerable methodological difficulties involved. In any case, dealing with these societal concerns can give rise to very large expenditures on safety measures, quite beyond the compass of the economically efficient decisions implied by the discretion given by Parliament to the regulators. In these circumstances, the process of negotiation between stakeholders and regulator, informed as best it is possible to do by characterisation and assessment of the risks and with final adjudication by the political system, must be paramount, as exhorted by Weinberg (1972) in the quotation given earlier.

In all aspects of the risk-based approach, there is much potential for scientific and technical contention as a result of uncertainties or disagreements about the applicability of data and information and in understanding of the processes which influence whether an undesired outcome is realised. It is therefore necessary that assumptions are made and judgements exercised by the protagonists in the technical debate. The mental processes involved and the scientific or engineering value systems applied in making those assumptions and judgements must necessarily be made transparent to the regulator and any others with an interest. The methodological way of achieving that transparency that has come to be accepted is the process of risk assessment.

12.8   Risk Assessment

Risk assessment is in essence a disciplined way of analysing and presenting predictions of the outcomes of activities or events that are subject to uncertainty. It consists of five essential stages (see, for example, HSE 1999b). These are:

- identification of the hazards or what is capable of going wrong. The process of identification may range from applying experience and information on known hazards in straightforward situations to the conduct of formal and structured methods in more complex and novel situations.
- identification of who might be harmed and how the harm might occur. The detail will again depend on the complexity of the situation and may involve the development of descriptive models of the various stages in the evolution of the chain of events leading to a harmful situation, the exposure of people to the consequences and the characterisation of the results of that exposure. The model descriptions may be in qualitative or quantitative terms, depending on whether a numerical estimate of the harmful effect is required.
- evaluation of the risks arising from the hazards and deciding whether the prevention, control and mitigation measures are adequate or whether more rigorous measures are necessary. Such decisions will be informed by reference to what the law requires and in particular the compliance or otherwise with standards of accepted and authoritative good practice in the relevant area of technology and safety management. The test is that any steps to reduce risks that are reasonably practicable in the circumstances should be implemented, giving the benefit of any doubt to the achievement of safety.

- the recording of the findings of the assessment and the reasons for the decisions made.
- the periodic review of the assessments in the light of relevant changes to circumstances and revision of the findings if necessary. Relevant changes would include the updating of what constitutes good practice and what is reasonably practicable in the light of advancing technology and the raising of standards to counter adverse experience or to respond to raised social expectations.

The degree to which quantification of the arguments is necessary or reliance placed on qualitative statements and judgements will vary with the nature of the hazard. For example, comparative risk assessment allows the risk of concern to be compared qualitatively with other known and adequately controlled risks to see whether it is higher or lower and, if higher, what then needs to be done to reduce the risk. It is essential, of course, that the risk is compared with risks of the same kind in terms of technical features and of the context in which the risk is being managed.

However the absence of genuine comparators and of relevant experience and proven good practice will enhance the value of quantification as an aid to the achievement of a convergence towards reasoned agreement. Such a situation arises where there is initially room for differing descriptions of the controlling phenomena. The increased application of quantitative methods in the risk assessment of a wide range of industrial hazards has been a notable feature of the past 25 years. As mentioned earlier, there are differing views on the value of quantification. To some, the quantitative approach has represented a desirable development, strengthening the rigour and objectivity of decision making. To others, it has made the process of decision making too exclusive, limiting the ability of non-experts to influence outcomes (Gerrard & Petts 1998), whilst the claim to objectivity has been challenged as being based on an unsustainable view of the nature of the probabilities which figure in the assessment (Watson 1994). The arguments about quantification of risk assessment and the complications that it brings have tended to overshadow the wider appreciation of the beneficial strengths of the risk-based approach to decision making. As further discussed below, quantification still features in these strengths, provided it is interpreted as subjective and to be used as a tool for dialectical debate (Watson 1994).

12.9   Strengths and Limitations of Risk Assessment

Good practice in risk assessment features a number of strengths which may be summarised as follows:

- it imposes *discipline* on the presentation of safety arguments by providing a structured framework to examine, display and debate all aspects of risks, their relative importance and their implications for risk control. This enables a robust demonstration to be given that all reasonably practicable steps to reduce risks have been rigorously examined.

- it helps to ensure *consistency*, i.e. a similar approach in similar circumstances, with attention focused on measures with the greatest effect, a result that may be easier to achieve on the basis of quantitative arguments.
- it provides *traceability and transparency* of the reasons for decisions on actions to be taken in relation to the assumptions on which the reasoning is based. This is particularly valuable in the subsequent monitoring of performance following implementation of the decision especially in respect of new evidence emerging which may challenge the validity of the assumptions.
- it enables actions to be made *proportionate* to the severity of the problem and clarifies the relative merits of different control strategies in terms of their effectiveness in reducing risk.
- it exposes the role of *judgement* at every stage in the weighing of arguments and enables differences between assessors to be pinpointed, analysed and resolved by constructive dialogue between the different interests.
- it ensures *compatibility* of intentions and assumptions in the assessment of the safety of a particular system with the actions of other players in the overall safety system.
- it provides a *source of defence* in the event of challenge.
- it enables *coherence* in the argumentation of safety cases such as are required in some areas of safety regulation.
- it allows the regulator to *control the delegation* of decisions where discretion to do so is permitted by the law.

These strengths are unfortunately accompanied by limitations on effectiveness which derive from the following features:

- it can be, and often is, misrepresented as an *objective* exercise whereas in fact it is pervaded by judgements about the underlying science.
- it is vulnerable to the influence of hidden *social value judgements* which should properly be reserved for consideration in the political or commercial decision on risk management.
- the complexities may make *communication* with the lay public difficult, especially since there is no universal agreement on language.
- an emphasis on *numerical estimates* of risk may result in lost contact with the public who assign importance to non-numerical risk characteristics, as listed in the earlier discussion on public perception of risks. The advocacy of a numerical scale of risk severity, analogous to the Richter scale for earthquakes, as a means of educating the public is seen as patronising.
- it concentrates on *individual* risk, whether qualitatively or quantitatively expressed. In many cases, this does not capture the concerns of the public. For example, these concerns may be more aroused by such matters

as the equitable distribution of risks and benefits, the manifestations of institutional failure, the imposition of unavoidable and irreversible limitations on freedom of choice and, increasingly prominent as a concern, the bequeathing of a legacy of problems to future generations.

12.10   Concluding Remarks

The discussion in this paper has focused on the factors that influence the response of individuals and society to risk issues. It is now widely recognised that these factors need to be taken into account in organising and presenting the scientific and technical input to the making of decisions on risk management. Some actions that are desirable in order to induce a climate in which constructive dialogue can take place on the handling of risk issues are as follows:

- the promotion of a greater understanding by the public of the processes of scientific enquiry and the limitations of those processes in providing objectively right answers to the archetypal question that concerns the public on risk issues, i.e. 'is it safe?';
- equally important, the promotion of a greater understanding by experts of what motivates public concerns and the consequent need for greater engagement with the public in the debates on risk issues;
- a concerted effort to inform the public about the elaborate safeguards provided by good practice in the conduct of risk assessment;
- where there is a conflict of views between experts, the development of a framework to get agreement on the differentiation of the aspects of the science pertaining to the risk issue that are not in contention from those that are and, in respect of the latter, to get agreement on what is necessary to arrive at the best available scientific advice.

A recent study (Hazards Forum 2000) is particularly relevant to identifying the way ahead in these matters. The study reports the results of consultation with many interests in the UK on measures that could be taken to improve the quality of public debates on risk issues. It provides suggestions on a number of fronts whose implementation, over time, should help to lessen contention on risk issues and hence the difficulties surrounding the making of workable and sustainable decisions on which economic progress and improvement of quality of life depend.

References

Adams, J. (1995). *Risk*, UCL Press, London.

Bacon, J. (1994). Risk-based regulation: setting goals for health and safety, PSAM-II Conference, 20-25 March, San Diego, CA.

Ball, D. J. & Floyd, P. J. (1998). *Societal Risks,* Report to Research Strategy Unit, HSE, Sheffield, UK.

Cabinet Office (2000). *Modernising Government Action Plan,* Cabinet Office, London.

Chilton, S. *et al.* (2000). *Valuation of Benefits of Health and Safety Control,* Report to Research Strategy Unit, HSE, Sheffield, UK.

Council for Science and Society (1977). *The Acceptability of Risks,* Barry Rose (Publishers) Ltd, London.

Department of Health (1999). *Saving Lives: Our Healthier Nation,* The Stationery Office, London.

Douglas, M. & Wildavsky, A. (1982). *Risk and Culture,* Univ of Calif Press, Berkeley, CA.

Douglas, M. (1994). Who is the Public?, in *Risks to the Public: The Rules, the Rulers and the Ruled,* Hazards Forum Symposium, 48-66, 14 Dec, London.

EPA (1997). *Framework for Environmental Risk Management,* Presidential/ Congressional Commission on Risk Assessment and Risk Management, Final Report, EPA, Washington DC.

Fischhoff, B., Slovic, P., Lichtenstein, S., Read, S., & Combs, B. (1978). How safe is safe enough? A psychometric study of attitudes towards technological risk and benefits, *Policy Studies,* **9**, 127-152.

Funtowicz, S. & Ravetz, J. R. (1990). Post-normal science: a new science for new times, *Scientific European,* 20-22, Oct.

Gerrard, S. & Petts, J. (1998). Isolation or integration? the relationship between risk assessment and risk management, in *Issues in Environmental Science and Technology* (Eds. Hester, R. E. & Harrison, R. M.), **9**, 1-19, Roy. Soc. Chem., Cambridge, UK.

Gibson, I. & Kass, G. (2001). A question of principle, science and public affairs, 14-16, April. Hazards Forum (2000), *Public Understanding of Risk,* London.

Health Council of the Netherlands (1996). *Risk is more than just a number,* Health Council of the Netherlands, The Hague.

Her Majesty's Treasury (1996). *The Setting of Safety Standards,* London.

House of Lords (2000). *Science and Society,* 3rd Report, Session 1999-2000, The Stationery Office, London.

HSE (1992). *The Tolerability of Risk from Nuclear Power Stations,* HSE Books, Sudbury, UK.

HSE (1999a). *Reducing Risks, Protecting People,* HSE Books, Sudbury, UK.

HSE (1999b). *Five Steps to Risk Assessment,* HSE Books, Sudbury, UK.

ILGRA (1996). *Use of Risk Assessment within Government Departments,* HSE Books, Sudbury, UK.

ILGRA (1998). *Risk Assessment and Risk Management,* HSE Books, Sudbury, UK.

Jones-Lee, M. W. (1994). Safety and the saving of life: the economics of safety and physical risk, in *Cost-Benefit Analysis, Second Edition,* (Eds. Layard, R. & Glaister, S.), 290-317, CUP, Cambridge, UK.

Kasperson, R. E., Renn, O., Slovic, P. & Brown, H. S. (1988). The social amplification of risk: a conceptual framework, *Risk Analysis,* **8,** 2, 177-187.

Laski, H. J. (1930). The limitations of the Expert, *Harper's Monthly Magazine,* December, 101-110.

Locke, J. H. (1981). *The Politics of Health and Safety,* Redgrave Memorial Lecture, Inst Occup Safety and Health, Leicester, UK.

MAFF (2000). *Procedures for Risk Analysis,* Ministry of Agriculture, Fisheries and Food, London.

MAFF (2001). *The Interim Response to the Report of the BSE Inquiry,* The Stationery Office, London.

McQuaid, J. (1997). Safety's debt to Davy and Faraday, *Proc. Roy. Inst. of Great Britain,* **68** , 177-208.

McQuaid, J. (2000). The application of risk control concepts to sustainable development, *Trans. I. Chem. E.,* **78,** B4, 262-269.

McQuaid, J. & Le Guen, J. M. (1998). The use of risk assessment in Government, in *Issues in Environmental Science and Technology* (Eds. Hester, R. E. & Harrison, R. M.), **9,** 21-36, Roy. Soc. Chem., Cambridge, UK.

OST (1997). *The Use of Scientific Advice in Policy Making,* Dept of Trade and Industry, London. (Revised 2000).

OXERA (2000). *Policy, Risk and Science: Securing and Using Scientific Advice,* Report to Research Strategy Unit, HSE, Sheffield, UK.

Petts, J., Horlick-Jones, T. & Murdock, G. (2001). Social amplification of risk: the media and the public, Report to Research Strategy Unit, HSE, Sheffield, UK.

Robens Committee (1972). *Safety and Health at Work,* HMSO, London.

Royal Commission on Environmental Pollution (2000). *Setting Environmental Standards,* 21st Report, The Stationery Office, London.

Royal Society (1992). *Risk: Analysis, Perception and Management,* Roy. Soc., London.

Royal Society (1997). *Science, Policy and Risk,* Roy. Soc., London.

Royal Society (1999). *Science, Technology and Social Responsibility,* Roy. Soc., London.

Starr, C. (1969). Social benefit versus technological risk, *Science,* **165,** 1232-1238.

Stern, P. C. & Fineberg, H. V. (Eds.) (1996). *Understanding Risk: Informing Decisions in a Democratic Society,* Nat. Acad. Press, Washington DC.

TRUSTNET (1999). *The TRUSTNET Framework: A New Perspective on Risk Governance,* TRUSTNET Secretariat, MUTADIS, 14 rue de Belzunce, 75010 Paris.

Waldegrave, W. (1987). *S*ustaining the Environment in a Developing World, NERC Annual Lecture, Roy. Soc., London.

Watson, S. R. (1994). The meaning of probability in probabilistic safety analysis, *Reliability Engineering and System Safety*, **45**, 261-269.

Weinberg, A. M. (1972). Science and Trans-Science, *Minerva*, **10**, 209-222.

Index